Soil Erosion - Current Challenges and Future Perspectives in a Changing World

Edited by António Vieira and Silvio Carlos Rodrigues

Published in London, United Kingdom

IntechOpen

Supporting open minds since 2005

Soil Erosion - Current Challenges and Future Perspectives in a Changing World
http://dx.doi.org/10.5772/intechopen.91595
Edited by António Vieira and Silvio Carlos Rodrigues

Contributors
Desale Kidane Kidane Asmamaw, Nurhussen Ahmed Mohammed, Andi Setyo Pambudi, Jules Rutebuka, Ahmed Benchettouh, Sihem Jebari, Lakhdar Kouri, Farris Okou, Achille Assogbadjo, Brice Augustin Sinsin, Henry Antonio Pacheco Gil, Argenis de Jesús Montilla Pacheco, Gayatri Sahu, Shreya Das, Samanyita Mohanty, Mausami Rana, Kiran Pilli

First published in London, United Kingdom, 2021 by IntechOpen
IntechOpen is the global imprint of INTECHOPEN LIMITED, registered in England and Wales, registration number: 11086078, 5 Princes Gate Court, London, SW7 2QJ, United Kingdom
Printed in Croatia

British Library Cataloguing-in-Publication Data
A catalogue record for this book is available from the British Library

Additional hard and PDF copies can be obtained from orders@intechopen.com

Soil Erosion - Current Challenges and Future Perspectives in a Changing World
Edited by António Vieira and Silvio Carlos Rodrigues
p. cm.
Print ISBN 978-1-83962-299-1
Online ISBN 978-1-83962-300-4
eBook (PDF) ISBN 978-1-83962-304-2

For EU product safety concerns:
IN TECH d.o.o., Prolaz Marije Krucifikse Kozulić 3, 51000 Rijeka, Croatia, info@intechopen.com or visit our website at intechopen.com.

Meet the editors

António Vieira is a geographer with a Ph.D. in Geography from the University of Coimbra, Portugal (2009). He is an assistant professor in the Department of Geography, University of Minho (UM), Portugal, and an integrated member of the Communication and Society Research Centre (CECS), UM. He is a member of several scientific organizations, including the Portuguese Association of Geomorphologists (Chair from 2017 to 2019) and the Portuguese Association of Risk, Prevention and Security (vice-president since 2015). He is also a member of FuegoRED and coordinator of the FESP International Network. Dr. Vieira's main research includes granitic geomorphology, geomorphological heritage and changes in land use, GIS and remote sensing, and their application to land use, geomorphological heritage, and soil erosion following forest fires and mitigation measures.

Silvio Carlos Rodrigues obtained a Ph.D. in Physical Geography from the University of São Paulo, Brazil, in 1998 and is a full professor at the Institute of Geography, Universidade Federal de Uberlândia (UFU), Brazil. He served as chair of the UGB in 2005–2006, secretary of the COMLAND of the IGA in 2010–2012, a full member of the CNPQ Advisory Committee on Physical Geography (CA-SA) in 2008–2010, and a full member of the Chamber of Natural Resources (CRA) with FAPEMIG in 2014–2016. Dr. Rodrigues is editor-in-chief of *Sociedade & Natureza* and a member of UNESCO's EOLSS International Editorial Council. He is also a reviewer for the *Brazilian Journal of Geomorphology*, *Caminhos de Geografia*, and *Environmental Earth Sciences*. He has experience in geosciences, with emphasis on geomorphology, acting mainly on the following themes: geomorphology, geomorphological cartography, soil erosion, integrated environmental analysis, and geomorphological mapping.

Contents

Preface **XI**

Section 1
Methodologies for Evaluation of Soil Erosion **1**

Chapter 1 **3**
RGB Spectral Indices for the Analysis of Soil Protection by Vegetation Cover against Erosive Processes
by Henry Antonio Pacheco Gil and Argenis de Jesús Montilla Pacheco

Chapter 2 **15**
Spatial Estimation of Soil Erosion Risk Using RUSLE/GIS Techniques and Practices Conservation Suggested for Reducing Soil Erosion in Wadi Mina Catchment (Northwest, Algeria)
by Ahmed Benchettouh, Sihem Jebari and Lakhdar Kouri

Chapter 3 **31**
Remote Sensing and GIS-Based Soil Loss Estimation Using RUSLE in Bahir Dar Zuria District, Ethiopia
by Nurhussen Ahmed Mohammed and Desale Kidane Asmamaw

Chapter 4 **57**
Determination of the Most Priority Conservation Areas Based on Population Pressure and Erosion Hazard Levels in Lesti Sub-Watershed, Malang Regency, Indonesia
by Andi Setyo Pambudi

Chapter 5 **73**
The Impacts of Soil Degradation Effects on Phytodiversity and Vegetation Structure on Atacora Mountain Chain in Benin (West Africa)
by Farris Okou, Achille Assogbadjo and Brice Augustin Sinsin

Section 2
Experiences on Soil Conservation **91**

Chapter 6 **93**
Erosion Control Success Stories and Challenges in the Context of Sustainable Landscape Management, Rwanda Experience
by Jules Rutebuka

Chapter 7 **121**
Biochar: A Sustainable Approach for Improving Soil Health and Environment
by Shreya Das, Samanyita Mohanty, Gayatri Sahu, Mausami Rana and Kiran Pilli

Preface

Soil is a fundamental component of Earth's environment. It is one of the subsystems of the Earth, being the interface between the hydrosphere, the atmosphere, and organisms that inhabit it. As its main functions, soil regulates natural material and energy cycles and is extremely sensitive to the effects of climate change and human and historical activities [1].

The genesis of soils is closely related to the changes undergone by the materials that make up the Earth's crust in proximity to the topographic surface. Subject to environmental conditions other than those that originated them, on the Earth's surface, rocks undergo a wide range of weathering processes, phenomena that essentially promote the breakdown of rocks, transforming them into detritic materials and chemical solutions.

Subjected to a diverse set of factors, soils undergo a degrading process, much of them of anthropic origin, which promote its erosion [2].

Soil erosion is an extremely serious environmental problem, widespread on practically the entire land surface, with direct and indirect effects on its productivity and thus human survival [3].

The consequences of this phenomenon are especially serious if we consider that the average rate of soil formation is around 1 t/ha/year, and the soil loss values are greater than 15 t/ha/year in China, 6 to 7 t/ha/year in the United States [4], and greater than 14 t/ha/year in Europe, in agricultural areas, vineyards, or soils without vegetation [5].

In addition, soil erosion can lead to the loss of 75% to 80% of its carbon content, causing the emission of carbon into the atmosphere [6].

It is a global problem and, although it is more serious in developing countries, it currently concerning in technologically more advanced countries.

Soil erosion is a phenomenon that occurs on practically the entire land surface, and in some areas, erosion and consequent deposition are essential for maintaining the soil's natural fertility.

However, the erosion action that occurs in the slopes also promotes the removal of the superficial part of the soil, precisely that where the highest concentration of nutrients is present [7]. When this process occurs at rates faster than those necessary for weathering and soil formation, its loss is irreversible.

If we realize that soil is responsible for 99% of the world's food production [8] then we will easily understand that it is critical to develop and implement soil erosion mitigation strategies and measures for protecting our soil while ensuring a sustainable and food-secure world.

Considering the importance and relevance of this topic for all societies, we have put forth much effort in developing this book. Written by authors from across the globe, the seven chapters in this volume reflect the authors' experiences of implementing different methodologies for soil erosion evaluation as well as soil conservation strategies.

Chapter 1 by Gil and Pacheco evaluates some RGB indexes for protecting the soil from erosive processes by vegetation cover, considering an area with agricultural use and based on images collected by Unmanned Aerial Vehicles (UVA). The objective of the proposed methodology, which was tested in Ecuador, is to offer the possibility of quickly differentiating vegetation from other types of coverage on the ground. The evaluation allowed to define which indexes present the best results and adaptation to the type of crop or plant mass mapped and to propose their use for zoning of risk of erosion under the agro-ecological conditions of the study area.

Chapters 2 and 3 propose the implementation of the Revised Universal Soil Loss Equation (RUSLE) to evaluate soil erosion in two African regions with different characteristics: Algeria and Ethiopia. In Chapter 2, Benchettouh et al. evaluate erosion in the wadi Mina catchment (Northwest Algeria) and its impact on silting up of dams built for human consumption and economic activities. The results of this study indicate that a significant part of the wadi Mina catchment (36.1%) is affected by high-to-dangerous erosion risk, revealing the urgent necessity of implementing soil conservation measures. In Chapter 3, Ahmed and Asmamaw propose the implementation of RUSLE to quantify the amount of soil loss in Bahir Dar Zuria district, Ethiopia, concluding the existence of a high correlation between soil loss and high slopes.

In Chapter 4, Pambudi presents a model for evaluating the Erosion Hazard Level (EHL) in the Lesti Sub-Watershed, Indonesia, considering population pressure to determine the priority conservation areas.

In Chapter 5, Okou et al. propose the evaluation of the impacts of soil degradation effects on phytodiversity and vegetation structure on the Atacora mountain chain in Benin, West Africa, concluding that physical soil degradation induced modification of floristic composition, phytodiversity loss and modification of vegetation structure.

In Chapter 6, Rutebuka presents some successful stories of sustainable landscape management and soil erosion control developed in Rwanda. These initiatives include intensive erosion control interventions as well as participatory landscape management, both promoted by the Government of Rwanda, to optimize land productivity in a sustainable manner.

In the final Chapter 7, Das et al. propose a revision of agronomic and environmental impacts of biochar on soil amendment, showing that biochar can play an important role in the modification of nutrients dynamics, soil contaminants, and microbial functions, which can give benefits to the soil, also strengthening soil erosion management.

António Vieira
CECS, Department of Geography,
University of Minho,
Guimarães, Portugal

Sílvio Carlos Rodrigues
Institute of Geography,
Federal University of Uberlândia,
Uberlândia, Brazil

References

[1] European Parliament (EP). 2002. European Parliament resolution on the Commission communication 'Towards a Thematic Strategy for Soil Protection'. COM(2002) 179 - C5-0328/2002 - 2002/2172(COS). https://eur-lex.europa.eu/legal-content/EN/TXT/?uri=uriserv%3AOJ.CE.2004.087.01.0070.01.ENG&toc=OJ%3AC%3A2004%3A087E%3ATOC#CE2004087EN.01039501

[2] Rodrigues, S.C.. Some Practices of Gully Rehabilitation in Central Brazil. In: Jagdish Chander Dagar,; Anil Kumar Singh. (Org.). Ravine Lands: Greening for Livelihood and Environmental Security. 1ed.Berlin: Springer Singapore, 2018, v. , p. 183-193.

[3] Vieira, A., Carvalho, I., Bento-Gonçalves, A., Rocha, J. (2020). Riscos geomorfológicos e suas manifestações. In L. Lourenço e A. Vieira (Coord.) Catástrofes Naturais. Uma abordagem global. Série Riscos e Catástrofes. Coimbra: Imprensa da Universidade de Coimbra, p. 119-167. https://doi.org/10.14195/978-989-26-1936-1_5

[4] Nearing, M. A., Xie, Y., Liu, B. and Yeb, Y. 2017. Natural and anthropogenic rates of soil erosion. International Soil and Water Conservation Research 5, 77–84.

[5] Boardman, J. and Poesen, J. (2006). Soil Erosion in Europe: Major Processes, Causes and Consequences. In J. Boardman and J. Poesen (Ed.), Soil Erosion in Europe. Chichester, England: John Wiley & Sons Ltd.

[6] Morgan, R. P. C. 2005. Soil Erosion and Conservation. 3rd edition. England: Blackwell Publishing.

[7] Bento-Gonçalves, A., Vieira, A., Coelho, C. and Ferreira, A. 2008. Caracterização geomorfológica e implementação de um sistema integrado de informação, em ambiente SIG, no âmbito do projecto RECOVER (estratégias de remediação de solos imediatamente após incêndios florestais). Revista Geografia Ensino & Pesquisa, 12(1), 3721–3735.

[8] Hatfield, J. L., Sauer, T. J. and Cruse, R. M. (2017). Chapter One - Soil: The Forgotten Piece of the Water, Food, Energy Nexus. Advances in Agronomy 143, 1-46.

Section 1

Methodologies for Evaluation of Soil Erosion

Chapter 1

RGB Spectral Indices for the Analysis of Soil Protection by Vegetation Cover against Erosive Processes

Henry Antonio Pacheco Gil
and Argenis de Jesús Montilla Pacheco

Abstract

The vegetation cover plays a fundamental role in protecting the soil from erosive processes. Many researchers have developed investigations for the calculation of the RUSLE C Factor, with the use of operating bands in the near infrared. With the current advances in Geospatial Technologies, there are a good number of RGB airborne sensors in Unmanned Aerial Vehicles (UVA). The objective of this chapter is to evaluate some RGB indexes, proposed in the literature, for the protection of the soil from erosive processes by vegetation cover, in a region with a high agricultural vocation. The methodology consisted of capturing RGB images in an area of the Ecuadorian coastal region and calculating in thematic indices, within the visible one, which offer the possibility of quickly differentiating vegetation from other types of coverage on the ground. The evaluation allowed to define which indexes present the best results and adaptation to the type of crop or plant mass mapped, and to propose their use for zoning of risk of erosion under the agro-ecological conditions of the study area.

Keywords: VIgreen, UAV, risk erosion, Manabí, Ecuador, RUSLE

1. Introduction

As is well known, vegetation cover plays a fundamental role in the protection of soil from erosive processes. Recent works [1–3] have developed research for the calculation of soil erosion with the Revised Universal Soil Loss Equation (RUSLE), using multispectral information in the visible bands and infrared mainly, for the analysis of vegetation cover. Additionally, [4, 5], used multispectral images of Landsat and Sentinel satellites to evaluate vegetation cover (C FACTOR RUSLE) through image classification processes, with spatial analysis tools for GIS software. The researchers conclude that the integration of remote sensors with GIS for the assessment of vegetation cover and soil properties represent suitable methods for forecasting changes in land use and accurately and easily measuring conditions that could lead to soil loss in the future.

With the current advances in Geospatial Technologies a number of airborne sensors are available in Unmanned Aerial Vehicles (UAVs) that dramatically improve the accuracy and resolution of information. Complex algorithms where used to detect topographical changes in agricultural surfaces with UAV images at different angles finding that vertical images are the most accurate to generate surface models that can be used in topographical evaluation, indispensable for the study of soil erosion [6]. On the other hand, the use of UAV has been reported to study the characteristics of the soil surface modified by the leveling, finding that these activities lead to a greater generation of runoff and sediment production [7].

There is a diversity of sensors for the use of UAVs in precision agriculture [8]. For the monitoring of crop and soil processes, the most commonly used sensors are those instrumented with multispectral cameras followed by thermal and hyperspectral camera and in the lastly RGB and infrared. Most works develop aerial monitoring processes that use machine learning or image processing techniques that include traditional indexes with multispectral bands.

The incorporation of multispectral sensors into UAV instruments, significantly increase costs, [9], and limit access to such technologies. For small producers with less economic resources it is proposed to use some alternative methods and indices, calculated and validated using conventional optics with visible bands [10–13]; accurate and reliable results are reported to analyze vegetation cover and its protective effect on soil. This possibility represents a strong competitive advantage for the processing of low-cost geospatial information relatively accessible to a larger number of users.

Permanent monitoring of vegetation cover is important to ensure sustainable management of agricultural activities, with a significant role in reducing water erosion. Beniaich used uncalibrated RGB images generated from a digital camera in an unmanned aerial vehicle (UAV), to assess 11 vegetation indices in the Bean and Mijo cycle study [11]. Vegetation indices with visible bands were effective tools for obtaining the soil coverage index compared to standard methods, resulting in these most practical and efficient rates in frequency and coverage area during the growing cycle.

In addition, orthomosaics in RGB have been used in multi-time periods to study physical and chemical characteristics of the soil. Better details were found for digital soil mapping, with multi-time-effective images and a classification overseen by the maximum likelihood method [5].

In summary, it shows the availability of a good variety of thematic indices within the visible range of the electromagnetic spectrum that offer the possibility of accurately and rapidly differentiating vegetation from another type of coverage on soil. For the analysis of this type of indexes it is very important to keep in mind that several of them do not respond to standardized formulas, therefore the resulting magnitudes can present a high fluctuation, and require processes of reclassification and interpretation of data according to each case.

2. Methodology

2.1 Study site

The study was developed in the experimental fields of the Faculty of Agricultural Engineering at the Universidad Tecnica de Manabí and Instituto Nacional de Investigaciones Agropecuarias (INIAP), Santa Ana town, Provinces of Manabí Ecuador (**Figure 1**). In the area of study there are different uses and soil cover, highlighting bare soils and permanent and short cycle experimental crops.

2.2 Flight platform and sensors

The aerial platform for image acquisition consisted of the UAV EBEE SQ. This SENSEFLY equipment is an advanced drone for agricultural use built around the

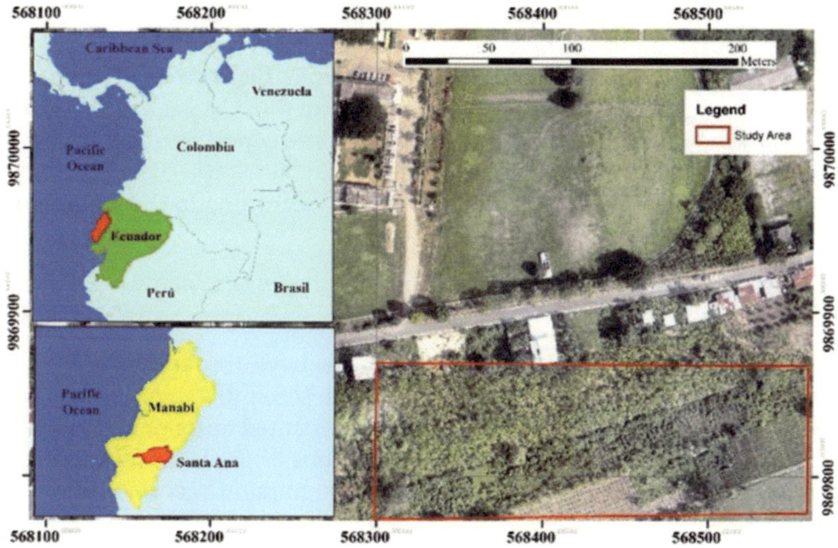

Figure 1.
Study area location.

revolutionary Parrot Sequoia camera. This multispectral sensor is a fully integrated solution in the UAV. Parrot Sequoia captures the Bands Red, Green, Red Edge (RE) and Near Infrared (NIR), and an RGB camera is also incorporated with the aim of generating an orthomosaic that supports crop analysis, as it takes the true colors of the terrain. The set is incorporated with a sunlight sensor that is located at the top of the EBEE which measures the intensity of sunlight at the time of the flight allowing normalization measurements made on different days with different light intensities, so multitemporal analyses can be performed on both cloudy and sunny days [14]. The RGB camera product was used for this research.

2.3 Imagery acquisition and processing

A photogrammetric flight was made for the acquisition of RGB images during the month of March 2020. The flight plan was programmed with the EMOTION AG software compatible with the UAV flight controller, for which the procedure specified below was fulfilled:

2.3.1 Physical inspection of the space to be flown

On the area selected for flight, the eye inspection of the space to be flown was carried out in order to identify physical elements that could cause an unwanted interruption to the flight of the drone, for example, very tall trees, antennas, sources of electromagnetism, buildings, etc. The inspection made it possible to clearly define the parameters to be considered in flight planning avoiding possible obstacles.

2.3.2 Planning the flight and photogrammetric support points

After the physical inspection the design of the flight was carried out, as well as the configuration of the shooting parameters, which are adapted to the different conditions of the flight plan, for example: the surface to be covered (which affects how quickly the photographs should be taken), flight height, transverse and longitudinal overlaps, flight speed, the flight's daylight (which affects the camera shutter parameters), among others.

For this phase, software compatible with the drone flight controller was used, in this case it corresponds to EMOTION AG incorporated in the package at the time of the equipment purchase. It is also possible to plan with open source software available to the entire interested community.

The flight was scheduled to run autonomously, according to the parameters specified in **Table 1**.

2.3.3 Flight execution

Once the flight plan was built, it was sent to the UAV through the Emotion Ag which communicates telemetrically with the drone through its receiving antenna. After the UAV recognized the flight plan, it was executed autonomously thanks to the team's geoinformatics equipment. The conditions of the UAV during the flight were permanently monitored in real time, until the mission was completed as planned.

2.3.4 Taking photogrammetric and post-process support points

For image processing the PIX4D application was used, it can also be any other post-processing software, with which the adjustments were made and verified the quality of the images acquired to obtain the quality report. For the entry of the photogrammetric support points, 10 control points were located on the ground. On each of these points the GNSS/GPS was positioned, accurately obtaining the coordinates that were entered in the processing of the images.

For the generation of the orthomosaic and RGB bands the photographs captured by the Parrot Sequoia camera were processed with the PIX4D software, which applies radiometric corrections that allow to normalize the images and compare different photographs on the same scale taking advantage of the advantage that this program and the multispectral camera Sequoia belong to the same company, [15]. Therefore, to calculate vegetative indexes these components are the most appropriate since they incorporate specific radiometric corrections, defined in the camera parameters and software processing algorithms.

This procedure ensured a high level of accuracy in the products generated (spectral bands), as input for the calculation of spectral indices with RGB information.

Information	Emotion AG
Camera	Multiespectral (1.2 Mpix)+RGI
Type	Sequoia 1.7.1
Image size (cm/pixel)	11.00 cm/px
Shutter time(s)	15:47 min
Flight area	24.2 ha
Longitudinal overlap (%) 80%	80%
Transverse overlap (%) 70%	70%
Flight height (m)	150 m
Flight speed (m s^{-1})	4 m/seg

Table 1.
Parameters obtained by the Emotion AG software for flight execution.

2.4 Vegetation fraction estimation

To estimate the plant fraction above soil, nine vegetation indices were evaluated, reported in recent literature as the most appropriate in terms of results and adaptation to the type of crop or plant mass mapped. The equations and fonts for each of the selected indexes are shown in **Table 2**.

Operations for calculating indexes, according to the equations listed in **Table 2**, were performed using the Spatial Analysis tools in ArcGIS software. Specifically

Index	Equation	Reported in
Vegetation Index Green	$\mathbf{VIgreen} = \frac{Gren-Red}{Green+Red}$	Costa et al. [10] Where r, g, and b are the normalized values of the bands R (Red), G (Green), and B (Blue), respectively
Visual Atmospheric Resistance Index	$\mathbf{VARI} = \frac{Gren-Red}{Green+Red-Blue}$	
Visible NDVI	$vNDVI = 0.5268 * (r^{-0.1294} * g^{0.3389} * b^{-0.3118})$	
Index Excess Green	ExG = 2g-r-b	Beniaich et al. [11] Where r, g, and b are the normalized** values of the bands R (Red), G (Green), and B (Blue), respectively
index Excess Green Minus Excess Red	EXGR = (2g-r-b) – (1.4r – g)	
Color Index of Vegetation	CIVE = 0.441r − 0.881g + 0.385b + 18.78745	
Soil Adjusted Vegetation Index	$\mathbf{SAVI} = \frac{1.5*(Gren-Rred)}{Green+Rred+0.5}$	Beniaich et al. [11]
Modified Green Red Vegetation Index	$\boldsymbol{MGVRI} = \frac{G^2-R^2}{G^2+R^2}$	Barbosa et al. [13]
Green Leaf Index	$\mathbf{GLI} = \frac{2G-R-B}{2G+R+B}$	

* * *The band values are transformed ranging from 0 to 1, according to: Marcial et al. [9]*
$\mathbf{Rn} = \frac{R}{R_{max}}\ \mathbf{Gn} = \frac{G}{G_{max}}\ \mathbf{Bn} = \frac{B}{B_{max}}$
where Rn, Gn, and Bn are the normalized values of its corresponding bands; R, G, and B are the original values of the red, green and blue, bands, respectively. Rmax = Gmax = Bmax are the maximum values for each band (255 for 24-bit colour images).
** *Obtaining the normalized spectral r, g, and b components, according to:*
$r = \frac{Rn}{Rn+Gn+Bn}\ g = \frac{Gn}{Rn+Gn+Bn}\ b = \frac{Bn}{Rn+Gn+Bn}$

Table 2.
Most appropriate RGB indexes for the plant study reported in recent specialized literature.

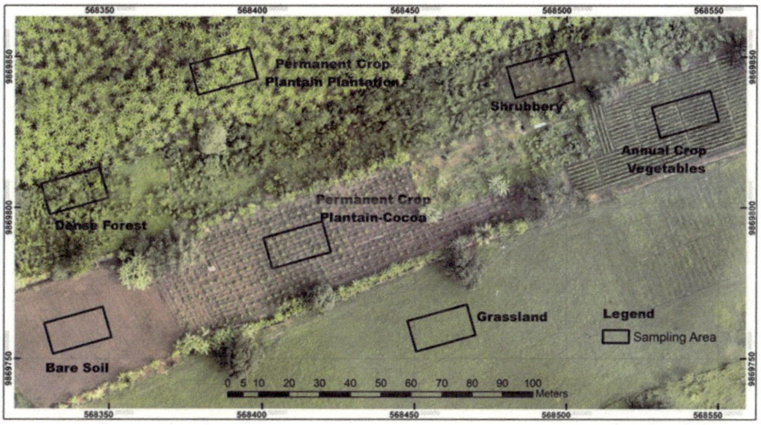

Figure 2.
Orthophoto with different coverages and selection of sampling areas.

worked on the Map Algebra tool, through which Spatial Analyst operators and functions were executed with the Raster Calculator, a simple Map Algebra expression was constructed and executed, using Python syntax in a calculator-like interface, designed for use in the application only as a tool dialog box [16].

2.5 Index evaluation according to coverage

To evaluate each of the indexes on the coverage selected in the orthophoto, **Figure 2**, a study area of 2.64 hectares was delimited with the presence of different land uses and cover. Using the principles of photointerpretation, 7 types of coverage on the area of interest were identified, which are shown in **Table 3**.

3. Analysis and results discussion

3.1 RGB image (orthophotos)

Figure 2 shows the true color composition (RGB), with excellent results in image reconstruction. Bare soil, in the southern west part of the image, is clearly differentiated, as well as early-stage crops in the central part and varied crops at different times of phenological development in the Eastern area. It stands out to the north, a banana crop in production stage with small spaces Inter crops without any protection to the soil.

The image shows very clearly the presence of a bare soil in preparation for cultivation, as well as an association of crops (banana and cocoa) in the initial state, where the development of the foliar area is still incipient and much of the soil is exposed to the watering processes. It is important to highlight that soil exposure to erosive processes, in this type of coverage should gradually decrease until very dense coverage in the medium term. It also highlights the presence of the low soil protection classes in the interplant areas of the Banana Monoculture, the alleys between the rows of the annual crop, and a specific area of fruit trees.

It is also evident specifically in areas of shrubs with the presence of tall grass and low-development trees, medium soil protection against erosive processes and a type of coverage, which represents the marginal areas of the different plant cover where good soil protection of erosive processes is initiated. This coverage is characterized by a perimeter appearance in permanent annual crops.

Good soil protection is characterized by the presence of dense vegetation and corresponds, in general, to foliar development in banana monoculture. Its expression is also very powerful in the branches of tall trees, represented in a rounded

Number	Coverage	Description
1	Null	Bare soil
2	Scarce	Permanent Crop (Plantain-Cocoa)
3	Median Low	Grassland
4	Medium dense	Annual cultivation (vegetables)
5	Dense	Shrubbery
6	Very Dense	Permanent Crop (Plantation)
7	Extremely dense	Dense forest

Table 3.
Types of coverage over the study area.

shape in the image. Likewise, they also represent this coverage in specific areas protected by small cocoa trees in the initial state of the associated crop.

There is a significant presence of two covers, with high capacity to protect the soil against erosive processes. For this particular case represented by dense forest with fruit and woodable species, as well as permanent crops in development stage and a small percentage on annual cultivation in maturity stage.

3.2 Spectral indexes

The indices obtained with RGB images were interpreted visually, grouping them into three classes (high, medium and low) according to their potential to discriminate different plant cover on the soil. **Figure 3** shows an example of each class.

Table 4 illustrates the classification of the different indexes into three categories according to their potential to discriminate covers, with different levels of soil protection before erosive processes.

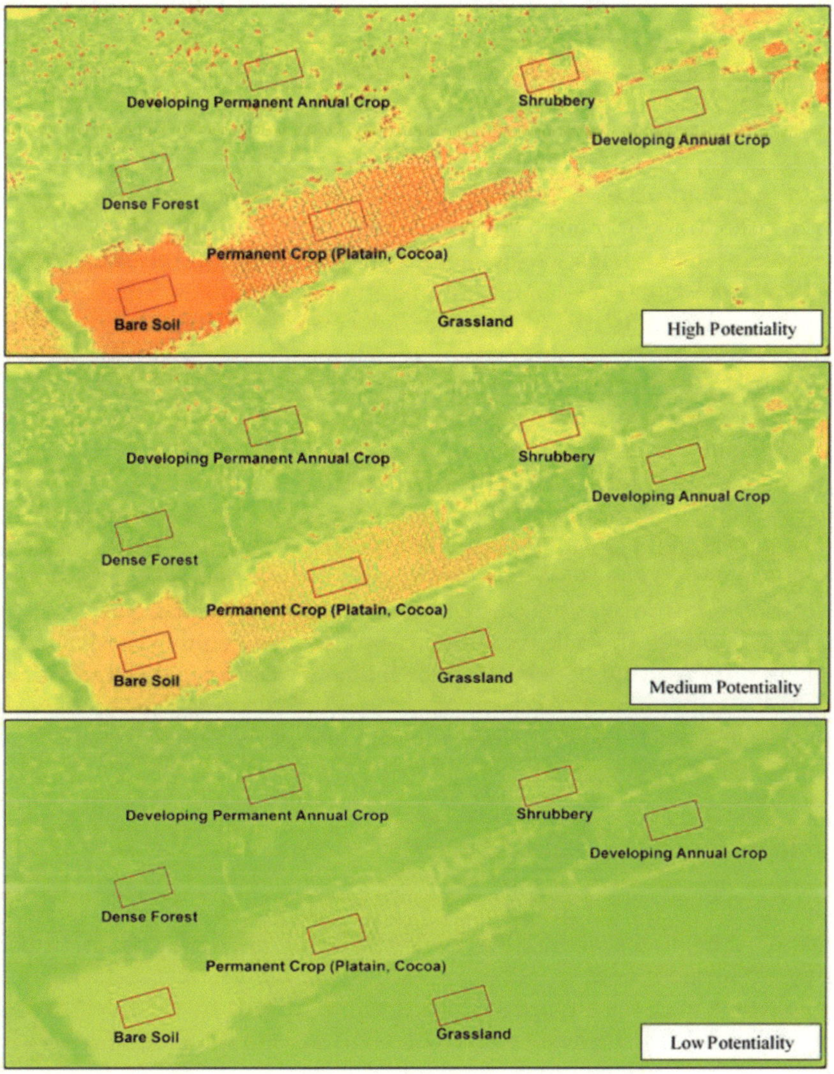

Figure 3.
Vegetation indexes with RGB data, classified according to their potential to discriminate different types of land cover.

Index	Equation	Categoría
Vegetation Index Green	$\mathbf{VIgreen} = \frac{Gren-Red}{Green+Red}$	High potentiality
Modified Green Red Vegetation Index	$MGVRI = \frac{G^2-R^2}{G^2+R^2}$	
Index Excess Green Minus Excess Red	$EXG = R(2g - r - b) - (1.4r - g)$	
Index Excess Green	$ExG = 2g - r - b$	Medium potentiality
Color Index of Vegetation	$CIVE = 0.441r - 0.881g + 0.385b + 18.78745$	
Soil Adjusted Vegetation Index	$\mathbf{SAVI} = \frac{1.5*(Gren-Rred)}{Green+Rred+0.5}$	
Green Leaf Index	$\mathbf{GLI} = \frac{2G-R-B}{2G+R+B}$	
Visual Atmospheric Resistance Index	$\mathbf{VARI} = \frac{\mathrm{Gren-Red}}{\mathrm{Green+Red-Blue}}$	Low potentiality
Visible NDVI	$vNDVI = 0.5268 * \left(r^{-0.1294} * g^{0.3389} * b^{-0.3118}\right)$	

Table 4.
Classification of spectral indexes according to their potential to identify different levels of soil protection.

The Vegetation Index Green (VIgreen), was ranked first regarding its potential to discriminate different ground covers. The Modified Green Red Vegetation Index and the Excess Green Minus Excess Red Index were placed in this same category. These indexes presented very similar results and could show differences, even in an apparently homogeneous area such as bare Soil coverage. Additionally, these indices showed very good ability to differentiate unprotected soil in interplant areas in a plantain crop. In the Image, the red colors represent bare soil with little to no plant protection and therefore a high risk of erosion.

With a medium capacity to differentiate some levels of soil protection, by vegetation cover, the Soil Adjusted Vegetation Index, Index Excess Green, Color Index of Vegetation, Green Leaf Index and Index Excess Green resulted. These indexes in general tend to slightly underestimate the areas with high erosion risks, in comparison with the previous indexes, especially in covers where the density of the vegetation is lower, as is the case of annual crops and shrubs.

On the other hand, the indexes with potential slime to zoning erosive risk in soil covers were the VARI and the vNDVI, which do not manage to differentiate exactly the part of the soil discovered between the crops and practically group the different types of soils in a heterogeneous class without plant protection.

4. Conclusions

Conventional optical information, with RGB images, allowed to generate very high quality orthophotos where different levels of soil protection against erosive processes can be identified.

The coverage with the highest soil protection is offered by the forest and permanent crops while the highest erosive risk was found in the ground covers in preparation and permanent crops in the initial stage.

Three of the calculated indexes offer a high potential to discriminate covers with different levels of soil protection, being VI green the one that showed the best performance, followed by MGVRI.

Four indexes were classified as medium potential and two as low, being the VARI and the vNDVI those that occupied the last places in terms of their potential, to analyze soil protection against processes.

Author details

Henry Antonio Pacheco Gil[1*] and Argenis de Jesús Montilla Pacheco[2]

1 Facultad de Ingeniería Agrícola, Universidad Técnica de Manabí, Portoviejo, Ecuador

2 Facultad de Hotelería y Turismo, Universidad Laica Eloy Alfaro de Manabí, Ecuador

*Address all correspondence to: henry.pacheco@utm.edu.ec

References

[1] Aneseyee AB, Elias E, Soromessa T, Feyisa GL. Land use/land cover change effect on soil erosion and sediment delivery in the Winike watershed, Omo Gibe Basin, Ethiopia. Sci Total Environ [Internet]. 2020;728:138776. Available from: doi:10.1016/j.scitotenv.2020.138776

[2] Eagleston H, Marion JL. Application of airborne LiDAR and GIS in modeling trail erosion along the Appalachian Trail in New Hampshire, USA. Landsc Urban Plan [Internet]. 2020;198(August 2019): 103765. Available from: doi:10.1016/j.landurbplan.2020.103765

[3] Phinzi K, Ngetar NS. The assessment of water-borne erosion at catchment level using GIS-based RUSLE and remote sensing: A review. Int Soil Water Conserv Res. 2019;7:27–46.

[4] Jazouli A El, Barakat A, Khellouk R, Rais J, Baghdadi M El. Remote sensing and GIS techniques for prediction of land use land cover change effects on soil erosion in the high basin of the Oum Er Rbia River (Morocco). Remote Sens Appl Soc Environ [Internet]. 2019;13: 361–74. Available from: doi:10.1016/j.rsase.2018.12.004

[5] Yang H, Zhang X, Xu M, Shao S, Wang X, Liu W, et al. Hyper-temporal remote sensing data in bare soil period and terrain attributes for digital soil mapping in the Black soil regions of China. Catena [Internet]. 2020;184 (September 2019):104259. Available from: doi:10.1016/j.catena.2019.104259

[6] Meinen BU, Robinson DT. Mapping erosion and deposition in an agricultural landscape: Optimization of UAV image acquisition schemes for SfM-MVS. Remote Sens Environ [Internet]. 2020; 239(January):111666. Available from: doi:10.1016/j.rse.2020.111666

[7] Peter KD, d'Oleire-Oltmanns S, Ries JB, Marzolff I, Ait Hssaine A. Soil erosion in gully catchments affected by land-levelling measures in the Souss Basin, Morocco, analysed by rainfall simulation and UAV remote sensing data. Catena [Internet]. 2014;113:24–40. Available from: doi:10.1016/j.catena.2013.09.004

[8] Radoglou-Grammatikis P, Sarigiannidis P, Lagkas T, Moscholios I. A compilation of UAV applications for precision agriculture. Comput Networks [Internet]. 2020;172(January):107148. Available from: doi:10.1016/j.comnet.2020.107148

[9] Marcial-Pablo M de J, Gonzalez-Sanchez A, Jimenez-Jimenez SI, Ontiveros-Capurata RE, Ojeda-Bustamante W. Estimation of vegetation fraction using RGB and multispectral images from UAV. Int J Remote Sens [Internet]. 2019;40(2):420–38. Available from: doi:10.1080/01431161.2018.1528017

[10] Costa L, Nunes L, Ampatzidis Y. A new visible band index (vNDVI) for estimating NDVI values on RGB images utilizing genetic algorithms. Comput Electron Agric [Internet]. 2020;172 (March):105334. Available from: doi: 10.1016/j.compag.2020.105334

[11] Beniaich A, Silva MLN, Avalos FAP, De Menezes MD, Cândido BM. Determination of vegetation cover index under different soil management systems of cover plants by using an unmanned aerial vehicle with an onboard digital photographic camera. Semin Agrar. 2019;40(1):49–66.

[12] Yeom J, Jung J, Chang A, Ashapure A, Maeda M, Maeda A, et al. Comparison of vegetation indices derived from UAV data for differentiation of tillage effects in agriculture. Remote Sens. 2019;11(13).

[13] Barbosa BDS, Ferraz GAS, Gonçalves LM, Marin DB, Maciel DT,

Ferraz PFP, et al. RGB vegetation indices applied to grass monitoring: A qualitative analysis. Agron Res. 2019;17 (2):349–57.

[14] SENSEFLY. eBeeSQ. 2020. https://www.sensefly.com/es/drone/ebee-sq/

[15] Zhou Y, Daakir M, Rupnik E, Pierrot-Deseilligny M. A two-step approach for the correction of rolling shutter distortion in UAV photogrammetry. ISPRS J Photogramm Remote Sens [Internet]. 2020;160 (November 2019):51–66. Available from: doi:10.1016/j.isprsjprs.2019.11.020

[16] ESRI. ¿Qué es el Álgebra de mapas? —ArcMap | Documentación [Internet]. 2020 [cited 2020 Sep 2]. Available from: https://desktop.arcgis.com/es/arcmap/latest/extensions/spatial-analyst/map-algebra/what-is-map-algebra.htm

Chapter 2

Spatial Estimation of Soil Erosion Risk Using RUSLE/GIS Techniques and Practices Conservation Suggested for Reducing Soil Erosion in Wadi Mina Catchment (Northwest, Algeria)

Ahmed Benchettouh, Sihem Jebari and Lakhdar Kouri

Abstract

To meet the pressing water needs in Algeria, the state has put in place a strategy consisting of the creation of hydraulic infrastructure for the mobilization of surface water resources. In fact, 74 dams are currently in operation; these structures are silting up at a rapid pace, generating an estimated annual loss of 45 million m^3. Sidi Mhamed Benaouda dam located in the Oranian hill, with a water capacity of respectively 241 million m^3 plays a crucial economic role in this region. The protection of this dam against erosive processes is a pressing economic goal. To do this, the RUSLE/GIS approach was used to map the erosive hazard. The results obtained in the Mina catchment, following a subdivision of 1315 homogeneous land parcels, show a total annual loss of 60 million tons/year with an average loss of 11.2 t/ha/year. About 50% of the catchment area was predicted to have very low to low erosion risk, with soil loss between 0 and 7.4 t/ha/year. Erosion risk is moderate over 13.9% of the catchment, where calculated soil loss is between 7.4 and 12 t/ha/year. Erosion risk is high to dangerous over 36.1% of the catchment, where calculated soil loss is more than 12 t/ha/year. According to this study, it appeared clearly that we must intervene quickly by using reliable and effective conservation techniques.

Keywords: Oranian hill, catchment, Sidi Mhamed Benaouda, soil loss modeling

1. Introduction

Water erosion is a phenomenon that results from the degradation of the surface layers of the ground cover and the displacement of the constituent materials [1] under the effect of the kinetic energy of the raindrops and the transport of soil particles from their original location [2]. It is one of the main causes of soil degradation in the world [3] leading to a significant threat to both human societies and the environment [4]. It also affects the quality of surface water and/or groundwater [5], reducing the capacity of the dams [6] and decreasing the soil fertility for agricultural activities [7]. Accordingly, the land area damaged by soil erosion is estimated at 1100 million hectares of land worldwide [8] resulting in the transportation of 2.0 to 2.5 x 10^{10} Mg of soil to the oceans each year [9]. This makes it a serious problem on a global scale and particularly worrying in certain regions of the world [10].

In 1930, in the United States, 20% of arable land was severely damaged by erosion following a prolonged drought. This is the dark era of the "dust bowl" phenomenon [11]. This resulted in the establishment of a water and soil conservation service by the US government. At the same time, a network of research stations was set up, which, thirty years later, resulted in the formulation of the USLE Equation [12].

Globally, [13] showed that of 13.5 billion hectares of land affected by water erosion, only 22% of the land is cultivable. During the last decades, the losses of cultivable land increased from 7 to 10 million hectares per year and at this rate, two centuries would be enough to destroy all the cultivable/agricultural land.

According to the United Nations Report on the State of the World's Soil Resources, published in 2015, cereal production losses due to erosion have been estimated at 7.6 million tons per year [14]. As a result of this report, researchers around the world have found that if nothing is done to mitigate erosion, we could achieve a reduction of more than 253 million tons of cereals by 2050. This loss of yield would be equivalent to removing nearly 15 billion hectares of land from farming [15]. According to this author, these more dramatic figures raised the alarm in different countries of the world in order to take all the necessary measures. In fact, due to the torrential nature of the rains, the high vulnerability of the land and the unfavorable human activities impact (deforestation, fires, overgrazing, poor agricultural behavior, chaotic town planning, etc.) more degradation will affect the agricultural landscape. Consequently, due to the relevance of this problem, several studies have been carried out on agricultural plots of about 100 square meters [2, 16–19], on micro-watersheds of a few hectares [20–23], on large basins of thousands of square kilometers [6, 24–26] and over large areas (countries and/or regions of the world) [27–29].

The results of soil loss vary from 1 to 200 t/ha/year (up to 700 t/ha/year) under crops specific to forest regions where slopes ranging from 30 to 60% and 0.5 to 40 t/ha/year under millet, sorghum, peanuts and cotton on long tropical ferruginous glacis of the Sudano-Sahelian regions whose slopes vary between 4 and 25% [30]. In the United States, on cultivated land, soil losses were estimated between 5 and 12 t/ha/year [31]. In Europe, [32] estimates that 25 million hectares have been seriously affected by erosion.

In the Maghreb, the water and soil potentials are seriously threatened [22, 33–36] and the phenomenon of water erosion is very widespread. The majority of watersheds are characterized by severe degradation exceeding 20 tons/ha/year [6], which leads to an average annual siltation of dam reservoirs at a rate of 125 million m^3 [37]. According to [38], water erosion in Morocco causes soil losses ranging from 5 t/ha/year to more than 50 t/ha/year depending on the region, and an average annual siltation of the reservoirs of the dams of the order of 75 million m^3. That is to say an annual reduction of 0.5% of their storage capacity, which causes deterioration in the quality of the drinking water mobilized and a decrease in water resources that can irrigate 10000 ha/year. Northern and central Tunisia currently has more than 30 dams with a total storage capacity of 3.5 billion m^3 [23, 38, 39]. Monitoring the siltation of these hydraulic structures made it possible to assess a loss of their storage capacity estimated at 30 million m^3/year, i.e. an annual reduction of 1%. Soil erosion has affected nearly 3 million hectares of agricultural land in the country, or more than half of the useful agricultural area in affecting the production capacity of Tunisian agriculture [23, 40].

In Algeria, the annual volume of sediment deposited in the 74 dams is estimated at 65 million m^3 [41]. Although soil erosion is characterized as a natural phenomenon, human activities such as agriculture can accelerate it further in Algeria [42]. Thus, 14 million hectares of land in the country are threatened by water erosion [43]. Therefore, Algeria is a country that witnesses an enormous deficit of water (i.e. below the theoretical scarcity threshold set by the World Bank, which is around 1000 m^3 per inhabitant/year) [44]. According to [45], Algeria is qualified in the category of the poorest African countries in terms of water potential. In 1962, the theoretical availability of water/capita/year was 1500 m^3; it was only 720 m^3 in 1990, 680 m^3 in 1995, 630 m^3 in 1998, 430 m^3 in 2020. To meet Algeria's urgent water needs, the States has implemented a strategy consisting of creating 94 hydraulic dams for the mobilization of surface water resources distributed throughout the national territory. The sector expected to build around 139 dams by 2030 [46]. One of these dams is that of wadi Mina catchment with a filling capacity of 241 million m^3. The dam Sidi Mhamed Benaouda (also named dam Es-Saada) is threatened by the silting from its site [6]. It is located at the extreme north of watershed of the wadi Mina (Algeria); its catchment area is subjected to intense water erosion with a volume of sediments which reaches the tank annually. This volume is on average about 3.2 million m^3 [47]. During these last decades and in a preoccupation with a management fight by the Algerian State, the catchment area of the wadi Mina was retained within the framework of a pilot project of integrated installation and development [47, 48]. The dam Sidi Mhamed Benaouda was built in 1978 with the downstream of this zone. According to [49], the marly sector located in the northern part constitutes the major source of sediments deposited in this dam.

Thus, the problem of water erosion mobilizes the scientific community to find solutions likely to ensure soil conservation [50]. In a context marked by global climate change and sustained human pressure on natural resources, the threat of soil erosion requires special and continued attention [51].

2. Methodology and data used

2.1 Tolerance to erosion

The tolerance level for soil loss varies from region to region of the world. It is linked to the productivity of the land and its uses [6]. Indeed, in Asia, [20] found that the tolerance threshold for soil loss in the Kelara sub-watershed in India was less than 1.5 t/ha/year. According to [52], the results of the study of water erosion in the Tamil Nadu basin (India) indicate that the average soil losses in this region are of the order of 6 t/ha/year. In Europe, [27] deduced that when soil losses exceed a threshold of 5 t/ha/year on cultivated land, the latter becomes intolerable. In a study conducted in the south-eastern region of Spain by [53], a rate of soil loss was recorded below the tolerable annual rates for the northern Mediterranean region. Almost 90% of its basins have average annual rates of less than 2 t/ha. In northwestern Turkey, the results obtained in the Buyukcekmece region by [54], show that a soil loss rate is low for a value of less than 1 t/ha/year, while beyond 10 t/ha/year, the phenomenon of erosion becomes a serious problem. Also, in the Alaca basin in Turkey, [55] estimated a water erosion tolerance rate of up to 12 t/ha/year. [56] underlines that any loss of soil greater than 1 t/ha/year is considered irreversible over a period of 50 to 100 years. A soil loss of 12 to 15 t/ha/year, or about 1 mm of soil per year (surface stripping) is sufficient to exceed the rate of alteration of the rocks. [54, 57] estimated the global tolerance rate for soil loss to be 10.2 t/ha/year.

In Morocco in regions similar to our study area, [58, 59] reported that soils can sustain loss of up to 7.4 t/ha/year on average. In Tunisia and according to the work of [22] carried out in the wadi Jannet watershed, a tolerance threshold of 8 t/ha/year has been suggested, above which the level of erosion risk will be high.

According to the above published work regarding the quantification of soil loss by the RUSLE model, it is clear that the tolerance threshold presents some difference. This is linked to the type of soil and its pedogenesis. In fact, in a region with shallow soil in a climate of accentuated summer aridity, the production of soil (pedogenesis) will be slow and consequently the tolerance threshold will be less. This is the case, for example, in arid and semi-arid Mediterranean regions [60].

2.2 Classification of soils and relationship with soil erosion

Given the objective pursued aimed at identifying the regions participating in the siltation of the Sidi Mhamed Benaouda dam, we have therefore adopted the American classification which is based on a tolerance threshold of 7.4 t/ha/year on average while supporting sustainably a high level of agricultural production and that if the losses exceed 20 t/ha/year, they can become dangerous [58]. We note that this classification has been adopted in Morocco, in the wadi Boussouab watershed, region similar to our study region in terms of climate, vegetation cover and soil substrate.

According to this classification, soil losses will be divided into five categories:

- Very low, when they do not exceed 5 t/ha/year.
- Low, when they are between 5 and 7.4 t/ha/year.
- Moderate when they are between 7.4 and 12 t/ha/year.
- Strong when they have values between 12 and 20 t/ha/year.
- Very strong and dangerous when losses exceed 20 t/ha/year.

2.3 Study area

Before proposing such a development, it is necessary to give a bibliographical overview of the research work relating to water erosion carried out in this region of interest. The watershed of wadi Mina is located at the northwest of Algeria in the Tellian hill area between 34° 42″ 36″ to 35° 35″ 2″ N latitude and between 0° 23″ 51″ to 1° 8″ 56″ E longitude (**Figure 1**). It lengthens on 90 km on Frenda and Mina mounts at north and on 50 km from the west to east between Bani-Chougrane mounts and the Ouarsenis massive. This watershed covers an area of 4800 km^2 [6].

2.4 Data used

The methodology adopted for this study rests on the exploitation of the multi-source data (satellite, pedological, climate condition and of the observations to carry out on the ground (in situ)). All these data are integrated and analyzed by the GIS for the cartography of the zones exposed to soil erosion in our study area.
1- Four images Shuttle Radar Topography Mission of resolution 30 m, coordinates: N35E000, N35E001, N34E000 and N34E001, were obtained from the SRTM (2011).
2- Two spectral scenes multi Landsat_8 OLI/TIRS (Operational Land To color) (Thermal Infrared Sensor) of Path_197/Row_035 (LC81970352014077LGN00) and

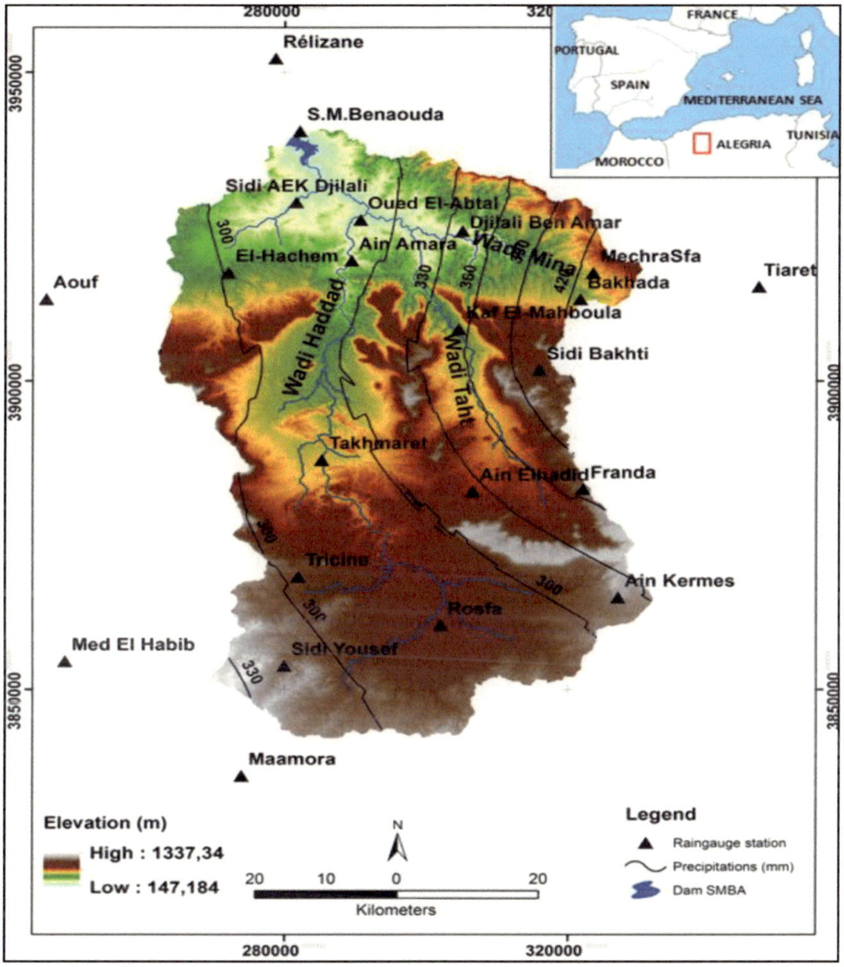

Figure 1.
Study area location map (source: [6]).

that of Path_197/Row_036 (LC81970362014077LGN00) were acquired on March 18th, 2014. These satellite images are uploaded from USGS (2014), with the Geotif format. 3- Observations on the ground, obtained after a descent on the ground in March and April 2014. 4- A detailed pedological map, catchment area of wadi Mina, drawn up by the BNEDER (2004). 5- Rainfall records (daily precipitations) provided by the National office of Meteorology on twenty two stations. They are spread over a period of 36 years (1978–2014) and cover our zone of interest.

2.5 RUSLE-Model

RUSLE- model proposes the same formula as the USLE [12] but several improvements were carried out for the determination of the various erosive factors. This included an approach different from the erodibility of the soil K-factor, a new equation for topographic LS-factor, and a new value for the crop management C-factor and the practices of conservation P-factor. The application of RUSLE model requires the calculation of the various factors intervening on the erosive processes and their spatialization in the form of the thematical maps. The integration of these data in the GIS makes it possible to superimpose them and evaluate the rate of water erosion by applying the formula of: $A = R*K*LS*C*P$.

Where: A: is the soil loss per unit of area (t/ha/yr). The R-factor is rainfall and runoff erosivity factor (MJ mm/ha h yr). K (t h/MJ mm) is soil erodibility factor, LS (unit-less) is a topographic factor, C (unit-less) is a crop management factor and P (unit-less) is a conservation practice factor.

3. Results and discussion

The combination of different thematic maps of erosive factors with their data bases was a subdivision of these into 1315 homogeneous plots with a total annual amount of land loss of 60 million tones. These losses vary between 0 t/ha/year and 521 t/ha/year, with an annual average of 11.2 t/ha and a standard deviation of 18.6/t/ha/yr. Spatially (**Figure 2**), the resulting map shows that the rate of soil loss varies from one sector to another in the study area. In fact, the low to very low soil loss classes are mainly located in the middle of the study area.

Although, this sector is characterized by steep topography and relatively high soil vulnerability. Our results are in agreement with those of the work of [61]. These authors found that in Algeria, not only runoff, but also soil erosion, does not systematically increase with the topography, in particular the slope. In addition, we note that the erosion risk in this sector is generally very low, recording an average of around 3.5 t/ha/year (**Table 2**). This clearly explains why the plant cover factor, in particular the forest one, plays a protective role. Indeed, [62] show that factor

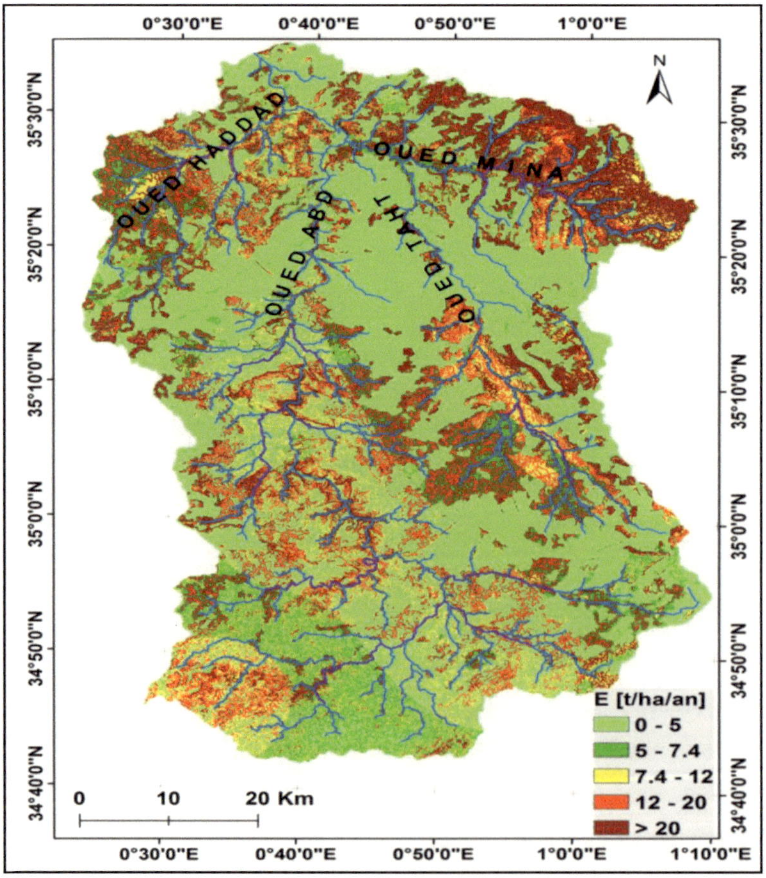

Figure 2.
Map of soil losses in the Wadi Mina basin.

C decreases the risk of erosion to 0.01 under perennial crops with cover plants or meadows and to 0.001 under forests associated with mulched crops compared to a bare plot.

The high and dangerous soil loss classes are noted exclusively in the northern and northeastern part of the study area. These regions are subject to an interweaving of natural and anthropogenic factors every year. The nature of the soils and the superficial lithological formation resulting from mainly marly terrain shows great fragility to water erosion. This is all the more important since the land has been almost completely bare and cultivated. [63, 64] show that these areas, which form an important part of the wadi Mina basin (1000 km^2), are strongly affected by water erosion.

According to the soil loss map obtained and according to the classification described above, the distribution of soil loss classes in the wadi Mina watershed is shown in **Table 1**.

The results of soil losses show that approximately 50% of the study area is classified in category where the erosion risk is low to very low (< 7.4 t/ha/year). 13.9% of the study area are classified in category where soil losses are moderate (7.4 to 12 t/ha/year). Actually, 36.1% of the study area is considered to be located in high risk and dangerous regions where losses exceed a threshold of 20 t/ha/year. The average rate of soil loss estimated at 11.2 t/ha/year is in the moderate erosion risk category.

Figure 3 highlights the following points (i) Sectors where soil loss exceeds the average of 11.2 t/ha/year, represent only 31.7% of the watershed. Their contribution to the overall soil loss is estimated at 92.5%. (ii) The sectors where the soil loss is lower than the average, occupy 68.3% of the surface of the basin. Their contribution represents only 7.5% of the global loss of soil.

The wadi Mina watershed has been the subject of several studies. These were carried out following the development of the Sidi Mhamed Benaouda dam and concerns raised by the scale of the erosive phenomenon and its consequences on the siltation

Risk of erosion	Soil loss class (t / ha / year)	Area (Km^2)	Area (%)	Category
Very low	0–5	1709.8	35.6	C
Low	5–7.4	688.9	14.4	
Moderate	7.4–12	665.9	13.9	B
Strong	12–20	903.7	18.8	A
Very strong and dangerous	> 20	831.7	17.3	

Table 1.
Soil loss classes in the Wadi Mina catchment.

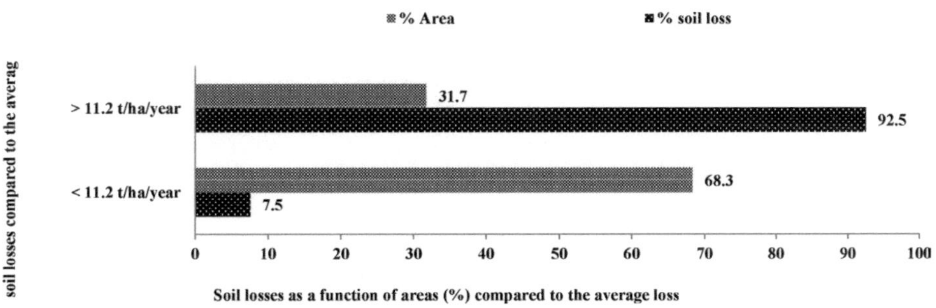

Figure 3.
Distribution of soil loss compared to the average in the Mina basin.

of the dam and the degradation of soil fertility from the 1990s. [65] established, from the classified parameters, the map of the sensitivity of marly lands to gullying in the western part of the watershed. This shows a predominance of land sensitive to linear erosion processes on the order of 57% of the territory. In fact, 25% of the land is strongly and very strongly sensitive to the incision and is mainly located on the right bank of wadi Mina, as well as in the downstream sector of the left bank of wadi Haddad. However, 18% of the basin surface is highly sensitive to solifluxion.

According to [62] and his collaborators observed that the different marly textures evolve by landslide and skin slide and those other environmental variables determine linear erosion namely: the slope, the vegetation cover and the morphology of the walls. The specific erosion of the wadi Mina basin estimated by the National Dams and Transfer Agency was on average around 3.26 t/ha/year, while the estimated soil losses in micro-watersheds with an area of 1000 km^2 located in the marly part can exceed a rate of 16 t/ha/year.

In 2001, [66] evaluated soil losses between 0.5 t/ha/year and 36 t/ha/year over the entire territory of the watershed. Most of this loss was recorded in the marl area with a rate exceeding 20 t/ha/year. However, [67] found that not only the marly areas participate in the production of sediments but the southern part of the basin of the wadi Mina can also participate with a significant contribution of sediments deposited in the lake of the Sidi Mhamed Benaouda dam.

In parallel, in 2004, the Algerian State under the supervision of Ministry of Agriculture and Rural Development launched a cooperation project with GTZ in order to develop a master plan for land use in the wadi Mina watershed. This is part of the conservation of soil and water strategy/planning. In 2006, under the supervision of the Ministry of Water Resources in collaboration with the Canadian consultancy firm (TECSULT), the Algerian State launched a study to identify and specify the measures to be undertaken to adequately fight against the siltation of reservoirs located in the Tellian hill including the wadi Mina basin which is one of these regions. The proposed developments have only been affected in areas classified as priorities A and B located in the marly region. According to this study, experts have shown that, if the improvements are made correctly in time, the lifespan of the Sidi Mhamed Benaouda dam will be increased twice as much as without it.

Before carrying out the development works according to the land use of our study area, it is first necessary to determine the average loss of each class as well as the degree of erosion risk. **Table 2** shows the sensitivity of the different types of land use to the risk of erosion. In fact, heavily vegetated areas, represented by vegetable crops and forests, are associated with low to very low soil losses, with 5.5 t/ha/year and 3.5 t/ha/year respectively. However, the higher and more dangerous ones correspond to bare soils with an average soil loss of around 29.8 t/ha/year. Soils used for agriculture, often protected during heavy spring showers, represent the type of vegetation cover most sensitive to erosion processes with an average soil loss of 16.1 t/ha/year. These last results are in agreement with those found by [68]. These authors have shown that cultivated fields can contribute significantly to sediment production. The formations based on scrub/scrubland, pasture and steppe produce moderate soil loss values with respective averages of 12 t/ha/year, 10.5 t/ha/year and 8.4 t/ha/year. This would be due to deforestation, overgrazing and bush fires which tend to substitute primitive formations for secondary cover of a different nature, such as savanna grassland.

The results of the evaluation of soil losses allowed us to deduce that an area of nearly 2400 km^2 of the slopes of the study area (**Table 2**) will require intervention measures to counter soil erosion. However, in order to optimize the allocation of resources intended for the short-term reduction of the siltation of the Sidi Mhamed Benaouda dam, we propose that only priority areas receive special attention in terms

DOI: http://dx.doi.org/10.5772/intechopen.96190

Land use class	Area		R (Mj.mm/ha.h.year)	K (t.h/Mj.mm)	LS	C*	E avrg. (t/ha/year)	Risk of erosion at the threshold of 7.4 t/ha/year
	Km^2	%						
Forest	821.0	17.1	864	0.144	2.7	0.01	3.5	Very low
Agglomerations	64.7	1.3	902.8	0.023	1.6	0	0	00
Agriculture	160.2	33.4	743	0.039	0.9	0.65	16.1	High
Firewall	75.4	1.6	872	0.013	1.9	1	1.9	High
Market gardening	718.4	15.0	844	0.203	1.8	0.018	5.5	Low
Scrub and Scrubland	532.6	11.1	889	0.034	1.8	0.25	12	Moderate
Pasture	61.3	1.3	489	0.19	1.03	0.1	10.5	Moderate
Bare soil	150.0	3.1	565	0.032	1.7	1	29.8	dangerous
Steppe	767.0	16.0	611	0.064	1.1	0.25	8.4	Moderate
Dam	8.4	0.2	0	0	0	0	0	00
Total/Moyenne	4800	100	523	0,0501	1,5	/	11,2	Modéré

C: Factor C data are obtained from National Research Institute for Rural Engineering, Water and Forestry (INRGREF), Tunisia (2014).*

Table 2.
Soil losses according to land use classes in the Wadi Mina basin.

of anti-erosion measures, including those classified in the two categories A and B where the risk of erosion is moderate to dangerous (> 7.4 t/ha/year) (**Table 1**).

Bare soils and firebreaks covering 150 km^2 and 75.4 km^2, respectively, are the main lithological occupations, and will require the most interventions in the watershed. These interventions are mainly intended to counter the gullying. These areas produce a significant amount of sediment estimated annually at an average of 29.8 t/ha and 19.9 t/ha respectively (**Table 2**). The protection strategy for these lands consists of installing torrential correction sills, constructing drains and outlets on slopes in order to avoid landslides with marly substrate, implanting dry stone lines and prohibiting their exploitation by livestock during the spring period when the soils must be covered.

The areas with agricultural activity adjacent to the Sidi Mhamed Benaouda dam and which come in second priority are responsible for a significant proportion of the siltation of this reservoir where soil losses exceed a threshold of 16.1 t/ha/year on an area of 1601.2 km^2, or 33.4% of the study area. The anti-erosion interventions recommended in these areas are the installation of arboriculture on terraces built along the contour lines, the establishment of living hedges, stone lines, drains and outlets along the road accesses as well as torrential correction thresholds in order to reduce the speed of runoff. In addition, it is important to ensure that farmers adhere to the principles of protection of arable slopes by mastering good mechanization which consists of working along the contour lines.

The scrubland and pastures with degraded soils and steppes are the third types of land use in terms of priority. Their erosion risk is between 7.4 t/ha/year and 12 t/ha/year requiring anti-erosion interventions. These interventions suggested over an area of approximately 600 km^2 include the establishment of torrential correction thresholds in the gullies, the planting of opuntia, revegetation and the installation of bulges, drains and outlets.

4. Conclusions

In view of our results, using the RUSLE approach in a GIS environment has many advantages, especially those related to the large number of findings. Indeed, it makes it possible to rationally manage a considerable quantity of quantitative and qualitative data relating to the various erosive factors. This allow, to disentangle their interdependence by successive crossing of thematic maps and to establish a synthetic map of the degree of erosion as well as the vulnerability of the different soils. Although the validity of soil losses is debatable, this method helps:

- Planners to suggest specific devices and techniques to prevent erosion processes

- Simulate landscape degradation while considering different management scenarios

Acknowledgements

The authors are very grateful to the Laboratory of Biodiversity, Conservation of Water and Soil (LBCWS) University of Abdel Hamid Ibn Badis, Mostaganem, to the University of Amar Telidji Laghouat (Algeria) and to the National Institute for Research in Rural Engineering, waters and forests (Tunisia). This scientific collaboration is dedicated to the exchange of experience and scientific know-how in the field of water resources management at the watershed scale. S. Jebari acknowledge helpful funding from the European Union Horizon 2020 programme, under Faster project, grant agreement N°[810812].

Author details

Ahmed Benchettouh[1*], Sihem Jebari[2] and Lakhdar Kouri[3]

1 Laboratory Biodiversity and Conservation of Water and the Soils, University Abdelhamid Ibn Badis, Amar Telidji University, Mostaganem, Laghouat, Algeria

2 National Research Institute for Rural Engineering, Water, and Forestry, Ariana, Tunis, Tunisia

3 Laboratory Biodiversity and Conservation of Water and the Soils, University Abdelhamid Ibn Badis, Kharouba, Mostaganem, Algeria

*Address all correspondence to: ahben17@yahoo.fr
and a.benchettouh@lagh-univ.dz

References

[1] Antoni, V., Le Bissonnais, Y., Thorette, J., Zaidi, N., Laroche, B., Barthès, S., ... & Arrouays, D. (2006). Modélisation de l'aléa érosif des sols en contexte méditerranéen à l'aide d'un Référentiel Régional Pédologique au 1/250.000 et confrontation aux enjeux locaux. Etude et Gestion des Sosls, *13*(3), 201-222.

[2] Kinnell, P.I.A. (2016). A review of the design and operation of runoff and soil loss plots. Catena 145 (2016) 257-265. 10.1016/j.catena.2016.06.013.

[3] Beatriz, S., Ranieri, L., Lier, Q. J., Sparovek, G., & Flanagan, D. C. (2002). Erosion database interface (EDI): A computer program for georeferenced application of erosion prediction models. Computer & Geoscience, 28(5), 661-668.

[4] Singh, R. P., Huerta-Espino, J., Sharma, R., Joshi, A. K., & Trethowan, R. (2007). Highyielding spring bread wheat germplasm for global irrigated and rainfed production systems. Euphytica, 157(3), 351-363.

[5] Ostovari, Y., Ghorbani-Dashtaki, S., Bahrami, H. A., Naderi, M., & Dematte, J. A. M. (2017). Soil loss estimation using RUSLE model, GIS and remote sensing techniques: A case study from the Dembecha Watershed, Northwestern Ethiopia. Geoderma regional, *11*, 28-36. 10.1016/j.geodrs.2017.06.003

[6] Benchettouh, A., Kouri, L., Jebari, S. (2017). Spatial estimation of soil erosion risk using RUSLE/GIS techniques and practices conservation suggested for reducing soil erosion in wadi Mina watershed (northwest, Algeria). Arab J Geosci,10:79. 10.1007/s12517-017-2875-6.

[7] Ahmed, K. R., & Akter, S. (2017). Analysis of landcover change in southwest Bengal delta due to floods by NDVI, NDWI and K-means cluster with Landsat multi-spectral surface reflectance satellite data. Remote Sensing Applications: Society and Environment, *8*, 168-181. 10.1016/j.rsase.2017.08.010

[8] Ganasri, B.P., Ramesh, H. (2015). Assessment of soil erosion by RUSLE model using remote sensing and GIS - A case study of Nethravathi Basin. Geoscience Frontiers, 1-9. 10.1016/j.gsf.2015.10.007.

[9] Markus, W. (2004). ECA annual report. *Eastern and Central Africa R Nairobi, Kenya.* http://www.worldagroforestry.org/eca/downloads/ECAAnnualReport.pdf.

[10] Mondal, S., Singh, R. P., Mason, E. R., Huerta-Espino, J., Autrique, E., & Joshi, A. K. (2016). Grain yield, adaptation and progress in breeding for early-maturing and heat-tolerant wheat lines in South Asia. Field Crops Research, *192*, 78-85. 10.1016/j.fcr.2016.04.017.

[11] Römkens, M. J. M. (2010). Erosion and sedimentation research in agricultural watersheds in the USA: from past to present and beyond. Sediment dynamics for a changing future, *337*, 17-26.

[12] Wischmeier, W. H. & Smith, D. D. (1978). *Predicting rainfall erosion losses: a guide to conservation planning* (No. 537). Department of Agriculture, Science and Education Administration.

[13] Kanwar, J. S. (1982). Managing soil resources to meet the challenges to mankind: presidential address.

[14] F.A.O. (2017). Les sols sont en danger, mais la dégradation n'est pas irréversible. Site web: http://www.fao.

org/news/story/fr/item/357221/icode. Consulté le, 27/01/2017.

[15] Benchettouh, A. (2019). L'érosion hydrique dans les régions marneuses du tell oranais cas des bassins versants de la Mina (Relizane) et Kramis (Mostaganem) (*Dissertation doctorale*). Université Abdelhamid Ibn Badis Mostaganem (Algérie)

[16] Bertol, I., Schick, J., Bandeira, D. H., Paz-Ferreiro, J., & Vázquez, E. V. (2017). Multifractal and joint multifractal analysis of water and soil losses from erosion plots: A case study under subtropical conditions in Santa Catarina highlands, Brazil. Geoderma, *287*, 116-125. 10.1016/j.geoderma.2016.08.008.

[17] Fernández, C., Vega, José, A. (2016). Evaluation of RUSLE and PESERA models for predicting soil erosion losses in the first year after wildfire in NW Spain. Geoderma 273, 64-72. 10.1016/j.geoderma.2016.03.016.

[18] Sadeghi, S.H.R., Sharifi, M. E., Khaledi D. A. (2016). Effects of subsequent rainfall events on runoff and soil erosion components from small plots treated by vinasse. Catena 138, 1-12. 10.1016/j.catena.2015.11.007.

[19] Anache, J.A.A., Wendland, E.C, Oliveira, P.T.S., Flanaganb, D.C., Nearing M.A. (2017). Runoff and soil erosion plot-scale studies under natural rainfall: A meta-analysis of the Brazilian experience. Catena 152, 29-39. 10.1016/j.catena.2017.01.003.

[20] Prasannakumar, V., Vijith, H., Abinod, S., & Geetha, N. J. G. F. (2012). Estimation of soil erosion risk within a small mountainous sub-watershed in Kerala, India, using Revised Universal Soil Loss Equation (RUSLE) and geo-information technology. Geoscience Frontiers, *3*(2), 209-215. 10.1016/j.gsf.2011.11.003.

[21] Ozsoy, G., Aksoy, E., Dirim, M. S., & Tumsavas, Z. (2012). Determination of soil erosion risk in the Mustafakemalpasa River Basin, Turkey, using the revised universal soil loss equation, geographic information system, and remote sensing. Environmental management, *50*(4), 679-694.

[22] Jebari, S., Berndtsson, R., Olsson, J., & Bahri, A. (2012). Soil erosion estimation based on rainfall disaggregation. Journal of hydrology, *436*, 102-110.

[23] Ben Slimane, A. (2013). *Rôle de l'érosion ravinaire dans l'envasement des retenues collinaires dans la Dorsale tunisienne et le Cap Bon* (Doctoral dissertation, Montpellier, SupAgro).

[24] Ganasri, B. P. & Ramesh, H. (2016). Assessment of soil erosion by RUSLE model using remote sensing and GIS-A case study of Nethravathi Basin. Geoscience Frontiers, *7*(6), 953-961. 10.1016/j.gsf.2015.10.007

[25] Haregeweyn, N. Tsunekawa, A., Poesen, J., Tsubo, M., Meshesha, D.T., Fenta, A.A., Nyssen, J. Adgo, E. (2017). Comprehensive assessment of soil erosion risk for better land use planning in river basins: Case study of the Upper Blue Nile River. Science of the Total Environment 574, 95-108. 10.1016/j.scitotenv.2016.09.019.

[26] Zhang, S., Fana, W., Li, Y., Yi, Y. (2017).The influence of changes in land use and landscape patterns on soil erosion in a watershed. Science of the Total Environment 574: 34-45. 10.1016/j.scitotenv.2016.09.024.

[27] Panagos, P., Borrelli, P., Meusburger, K., Van der Zanden, E. H., Poesen, J., Alewell, C. (2015). Modelling the effect of support practices (P-factor) on the reduction of soil erosion by water at European scale. Environnemental science et policy 51: 23-34. 10.1016/j.envsci.2015.03.012.

[28] Zhang, W., Zhou, J., Feng, G., Weindorf, D. C., Hu, G., Sheng, J. (2015). Characteristics of water erosion and conservation practice in arid regions of central Asia: Xinjiang province, China as an example. International soil and water conservation reseach.2095-6339, 30061-7. 10.1016/j.iswcr.2015.06.002.

[29] Wang, J., Huang, J., Zhao, X., Wu, P., Horwath, W.R., Li, H., Jing, Z., Chen, X. (2016). Simulated study on effects of ground managements on soil water and available nutrients in jujube orchards. Land Degradation & Development 27, 35-42. 10.1002/ldr.2334.

[30] BOLI, B. Z., Roose, E., Ziem, B. B. A., Sanon, K., & Waechter, F. (1993). Effets des techniques culturales sur le ruissellement, l'érosion et la production de coton et maïs sur un sol ferrugineux tropical sableux. Cahier ORSTOM, Série Pédagogique, *28*(2), 309-325.

[31] Pimentel, D., Harvey, C., Resosudarmo, K., Sinclair, K., Kurz, D., McNair, M., Crist, S., Shpritz, L., Fitton, L., Saffouri, R., Blair, R. (1995). Environmental and economic costs of soil erosion and conservation benefits. Science 267, 1117-1123.

[32] De Ploey, J. (1989). A Soil Erosion Map for Western Europe. Catena Verlag.

[33] Roose, E., Arabi, M., Brahmia, K., Chebani, R., Mazour, M., Morsli, B. (1993). Recherche sur la réduction des risques d'érosion par la GCES en moyenne montagne méditerranéenne algérienne. *Cahier ORSTOM*, Sérier Pédologique, 28 (2), 289-307.

[34] Touaibia, B. (2000). Erosion-Transport Solide-Envasement de barrage.Cas du bassin versant de l'Oued Mina - Wilaya de Relizane. Thèse de doctorat d'Etat, Institut Nationale Agronomique, Alger, 200 p.

[35] Mazour, M. et Roose, E. (2002). Influence de la couverture végétale sur le ruissellement et l'érosion des sols sur parcelles d'érosion dans les bassins versants du nord-ouest de l'Algérie Bull Réseau Erosion; 21 ·320-30.

[36] Morsli, B., Habi, M., Mazour, M., Hamoudi, A., Halitim, A. (2012). Erosion et ruissellement en montagnes méditerranéennes d'Algérie du Nord: analyse des facteurs conditionnels sous pluies naturelles et artificielles. *Rev. Mar. Sci. Agron. Vét.*1:33-40

[37] Remini, W., Remini, B. (2003). La sedimentation dans les barrages de l'Afrique du nord. Larhyss Journal, ISSN 1112-3680, n° 02, Juin 2003, pp. 45-54.

[38] Souadi, Y. (2011). L'érosion hydrique au maghreb étude d'un cas: le bassin" versant de l'oued barbara (tunisie septentrionale). Mémoire présenté comme exigence partielle de la maitrise en géographie. Université du Québec à Montréal. 143 P.

[39] Hajji, O., Abidi, S, Hermassi, T., Habaieb, H., Raouf, M.M. (2015). Cartographie de la vulnérabilité multifactorielle à l'envasement des lacs collinaires de la Tunisie centrale. International Journal of Innovation and Applied Studies ISSN 2028-9324 Vol. 10 No. 4, pp. 1251-1266.

[40] Jebari, S. (2009).Water Erosion Modeling using Fractal Rainfall Disaggregation – A Study in Semiarid Tunisia. PhD Thesis, Rep. 1047, Dept. of Water Resour. Engng., Lund Univ, pp. 1-140.

[41] Remini, B. Une nouvelle approche de gestion de l'envasement des barrages. Larhyss Journal, (31) (2017) 51-81

[42] Mahé, G., AlKarkouri, J., Benabdelfadel, H., Bensafia, D., Brou, T., Chaker, M., Chikhaoui, M., Coupleux, S., Crouzevialle, R., Dieulin, C., Emran, A., Ezzaouini, M.,

Goussot, E., Hallouz, F., Khomsi, K ., Laouina, A., Machouri, N., Maleval, V., Meddi, M., Nging, M., Planchon, O., Remini, B., Saadi, H., Sfa, M., Sinan, M., Snoussi, M., Taibi, S., Toumi, S., Tra Bi, A. (2013). Etude de l'évolution de l'occupation du sol sur deux grands bassins d'Algérie et du Maroc, et relation avec la sédimentation dans les barrages. In: Considering hydrological change in reservoir planning and management (Schumann, A. *et al.*, Eds.) IAHS Publ. 362, 115-123.

[43] Meddi, M., Toumi, S., Assani, A.A. (2016). Spatial and temporal variability of the rainfall erosivity factor in Northern Algeria. Arabian Journal of Geosciences, 9(4) 282. 10.1007/s12517-015-2303-8

[44] Bouzid, T. (2010). Les barrages et la politique hydraulique en Algérie: état, diagnostic et perspectives d'un aménagement durable (Dissertation doctorale). Université de Mentouri–Constantine.

[45] Mohammed, T., Al-Amin, A.Q. (2018). Climate change and water resources in Algeria: vulnerability, impact and adaptation strategy. Economic and Environmental Studies, 18(1) (2018) 411-429. 10.25167/ees.2018.45.23

[46] Ministry of Water Resources and Environment (MWRE) (2020). Algeria - Laghouat | The project of completing the Sklava Dam with a capacity of 55 million cubic meters. Retrieved from https://www.youtube.com/watch?v=h_xYEv8P1bI

[47] Achite, M., Touaibia, B., & Ouillon, S. (2006). Erosion hydrique en Algérie du Nord: Ampleur, conséquences et Perspectives. In *14th International Soil Conservation Organization Conference. Water Management and Soil Conservation in Semi-Arid Environments. Marrakech, Morocco*.

[48] GTZ (1996). L'aménagement des zones Marneuses dans les bassins-Versants des Montagnes de l'Atlas Tellien Semi-Aride, Deutsche Gesellschaft fur Technische Zusammenabeit (GTZ).

[49] Kouri, L., & Gomer, D. (2010). Analyse des processus de ravinement et détermination de la sensibilité des terrains marneux dans le Tell oranais (Algérie). Science et changements planétaires/Sécheresse, *21*(4), 257-265.

[50] Aké, G. É., Kouadio, B. H., Adja, M. G., Ettien, J. B., Effebi, K. R., & Biémi, J. (2012). Cartographie de la vulnérabilité multifactorielle à l'érosion hydrique des sols de la région de Bonoua (Sud-Est de la Côte d'Ivoire). *Physio Géo. Géographie physique et environnement*, (Volume 6), 1-42.

[51] Aiello, A., Adamo, M., Canora, F. (2015). Remote sensing and GIS to assess soil erosion with RUSLE3D and USPED at river basin scale in southern Italy. Catena 131, 174-185. 10.1016/j.catena.2015.04.003.

[52] Balasubramani, K., Veena, M., Kumaraswamy, K., & Saravanabavan, V. (2015). Estimation of soil erosion in a semi-arid watershed of Tamil Nadu (India) using revised universal soil loss equation (rusle) model through GIS. Modeling Earth Systems and Environment, *1*(3), 10. 10.1007/s40808-015-0015-4.

[53] Sougnez, N., van Wesemael, B., & Vanacker, V. (2011). Low erosion rates measured for steep, sparsely vegetated catchments in southeast Spain. Catena, *84*(1-2), 1-11.

[54] Demirci, A., & Karaburun, A. (2012). Estimation of soil erosion using RUSLE in a GIS framework: a case study in the Buyukcekmece Lake watershed, northwest Turkey. Environmental Earth Sciences, *66*(3), 903-913.

[55] Imamoglu, A., Dengiz, O. (2016). Determination of soil erosion risk using RUSLE model and soil organic carbon loss in Alaca catchment (Central Black Sea region, Turkey). Rend. Fis. Acc. Lincei.: 10.1007/s12210-016-0556-0.

[56] Kouli, M., Soupios, P., & Vallianatos, F. (2009). Soil erosion prediction using the revised universal soil loss equation (RUSLE) in a GIS framework, Chania, Northwestern Crete, Greece. Environmental Geology, *57*(3), 483-497.

[57] Yang, D., Kanae, S., Oki, T., Koike, T., & Musiake, K. (2003). Global potential soil erosion with reference to land use and climate changes. Hydrological processes, *17*(14), 2913-2928.

[58] Sadiki, A. (2004). Estimation des taux d'érosion et de l'état de dégradation des sols dans le bassin versant de l'oued Boussouab, Maroc nord oriental : application du modèle empirique (USLE), de la technique du radio-isotope ^{137}Cs et de la susceptibilité magnétique. Thèse Doctorat d'Etat. Université Mohamed Premier, Oujda, 320 p.

[59] Yjjou, M., Bouabid, R., El Hmaidi, A., Essahlaoui, A., & El Abassi, M. (2014). Modélisation de l'érosion hydrique via les SIG et l'équation universelle des pertes en sol au niveau du bassin versant de l'Oum Er-Rbia. The International Journal Of Engineering And Science (IJES), *3*(8), 83-91.

[60] Kheir, R. B., Girard, M. C., Khawle, M., & Abdallah, C. (2001). Érosion hydrique des sols dans les milieux méditerranéens. Études et gestion des sols, *8*(4), 231-245.

[61] Kouri, L. (1993). *L'erosion hydrique des sols dans le bassin versant de l'oued mina (algerie). Etude des processus et types fonctionnels de ravins dans la zone des marnes tertiaires* (Doctoral dissertation, Strasbourg 1).

[62] Roose, É., Sabir, M., Arabi, M., Morsli, B., & Mazour, M. (2012). Soixante années de recherches en coopération sur l'érosion hydrique et la lutte antiérosive au Maghreb. *Physio-Géo. Géographie physique et environnement*, (Volume 6), 43-69.

[63] Toumi, S. (2013). *Application des techniques nucléaires et de la télédétection à l'étude de l'érosion hydrique dans le bassin versant de l'oued Mina* (Doctoral dissertation, Thèse de Doctorat, Ecole Nationale de l'Hydraulique, Algérie, p 9 et 36).

[64] Achite, M., & Meddi, M. (2004). Estimation du transport solide dans le bassin-versant de l'oued Haddad (Nord-Ouest algérien). Science et changements planétaires/Sécheresse, *15*(4), 367-373.

[65] Achite, M., & Ouillon, S. (2007). Suspended sediment transport in a semiarid watershed, Wadi Abd, Algeria (1973-1995). Journal of hydrology, *343*(3-4), 187-202.

[66] Touaibia, B., Gomer, D., Aïdaoui, A. et Achite, M. (2001). Quantification et variabilité temporelles de l'écoulement solide en zone semi-aride, de l'Algérie du Nord. Journal des sciences hydrologiques 46 (1), Pp. 41-53.

[67] Touaibia, B., Bouaïchi, I., & Dernouni, F. (2006). approche de calcul de l'exposant climatique en cas d'insuffisance de donnees: cas de la region de Tipaza. Production scientifique-Communications, 1.

[68] Alejandro, M., & Omasa, K. (2007). Estimation of vegetation parameter for modeling soil erosion using linear Spectral Mixture Analysis of Landsat ETM data. ISPRS Journal of Photogrammetry and Remote Sensing, *62*(4), 309-324.

Chapter 3

Remote Sensing and GIS-Based Soil Loss Estimation Using RUSLE in Bahir Dar Zuria District, Ethiopia

Nurhussen Ahmed Mohammed and Desale Kidane Asmamaw

Abstract

The severity of soil loss in the Ethiopian highlands has been increased from time to time. Hence, the assessment of soil erosion using models is very important for planning successful and sustainable soil management. This study was conducted in Bahir Dar Zuria district, Ethiopia with aiming to quantify the amount of soil loss using the GIS-based RUSLE (Revised Universal Soil Loss Equation) model. Based on the study, the most pronounced RUSLE factor that increases soil erosion was the slope length (L) and slope steepness (S). Compared with other land uses, bare land and cropland in the higher slopes were more vulnerable to erosion. As expected slope and soil losses have a direct relationship. About 80% of the study area experienced annual soil loss of less than 1.2 ton/ha/yr. Conversely, soil loss was very high for slopes greater than 30%. This indicated that slope has a great impact on regulating soil loss. The annual soil loss for cropland, vegetation, grassland, and degraded land was 19.05, 8.78, 8.82, and 71.16 ton/ha/yr., respectively. This is to means that land use land cover have a strong relationship with the amount of soil loss. The same land cover with different slopes have different soil loss amount. It was found that lack of vegetative cover during the critical period of rainfall, expansion of croplands, and absence of support practices increase soil erosion. Thus, the application of stone lines, contour tillage, terraces, and grass strip barriers are suggested to break the slope length into shorter distances, reducing overland flow velocity and soil erosion. Moreover, improving the awareness of society to reduce the illegal cutting of trees and apply conservation practices to reduce soil erosion in their farmland is very essential.

Keywords: Ethiopia, GIS, land use cover change, RUSLE, soil loss, slope

1. Introduction

Land degradation has been one of the major global environment and sustainable development challenges in the 21st Century. The expansion of agriculture and the clearance of natural habitats over the past decades aggravated the magnitude of land degradation in Ethiopia [1, 2]. Land degradation is mainly manifested by soil erosion [3].

Soil erosion is a serious problem in the Ethiopian highlands that increased sedimentation of reservoirs and lakes [4, 5]. Sediment export rates in the Ethiopian highlands are characterized by important changes in sediment supply [2, 6–9]. FAO [10] reported that soil erosion in Ethiopia is nearly 10 times greater than the rate of soil regeneration, and the country has among the highest estimated rates of soil nutrient depletion in Sub-Saharan Africa. Such land degradation reduces average agricultural productivity. It also increases farmers' vulnerability to drought by reducing soil fertility and water-holding capacity. Thus, land degradation in the form of soil erosion and declining soil quality is a serious challenge to agricultural productivity and economic growth in these highlands [11].

Soil erosion is a hazard traditionally associated with agriculture in different parts of the world and is important for its long-term effects on soil productivity and sustainable agriculture [1, 5, 11]. It is, however, a problem of wider significance occurring additionally on land devoted to forestry, transport, and recreation. Hence, it is important to identify estimated locations where soil erosion occurs to prevent substantial soil loss. In the most erodible situations, soil loss or sediment yield is limited by the transport capacity of the runoff. As the runoff flows through a watershed, changes in topography, vegetation, and soil characteristics often reduce this transport capacity [12–15].

Severe soil erosion not only leads to the impoverishment of cultivated land and poverty of the local people, but also to desertification that destroys the conditions crucial for human survival. It leads to the reduction of land/soil quality, loss of topsoil, and decrease in the content of soil organic matter and thereby to the loss in crop yield as it relates to high runoff rates and low soil permeability which in turn resulted in a decrease in infiltration and less water availability for the crops [16].

Degradation of land indicates undesirable changes that destroy the potentialities of regeneration, growth, and survival of plants. It is one of the most serious environmental problems causing great concern. Degradation is a cumulative effect of various factors acting singly or in combination [1]. Addressing land degradation would, therefore, could contribute significantly to reducing poverty and ensuring environmental sustainability.

The importance of studying soil erosion among global issues is enhanced because of its impact on world food security and the quality of the environment. The severity of the land degradation process makes large areas unsuitable for agricultural production because of the removal of topsoil and even part of the subsoil in some areas, and stones or bare rock are left at the surface [17]. Thus, there is a growing global awareness that land degradation is as much a threat to environmental well–being as more obvious forms of damage, such as air and water pollution.

To restore the productivity of the soil and to prevent further damage, planning, conservation, and management of the watersheds are vital. The watershed prioritization and formulation of proper watershed management programs for sustainable development require information on watershed sediment yield [18]. However, due to the complexity of the variables involved in the erosion process, it becomes difficult to measure or predict the soil loss in a precise manner [19]. Conversely, remote sensing data provide accurately, and near-real-time information on the various aspects of the watershed such as land use/land cover, physiographic, soil distribution, drainage characteristics, etc. [19]. It also assists in the identification of the existing or potential erosion-prone areas and provides data inputs to many of the soil erosion and runoff models [20].

To quantify the sediment yield (soil loss), several empirical models based on the biophysical parameters were developed in the past [7]. Among other models, Sediment Yield Index (SYI) [17] and Universal Soil Loss Equation (USLE) [10]

DOI: http://dx.doi.org/10.5772/intechopen.95393

are extensively used. For instance, the USLE model has been widely applied at the watershed scale based on the lumped approach [17, 21] to the catchment scale [21]. However, various modifications in the models were often applied for the estimation of soil loss using GIS and remote sensing [22]. The Revised Universal Soil Loss Equation (RUSLE) uses the same empirical principles as USLE, however, it includes numerous improvements, such as monthly factors, incorporation of the influence of profile convexity/concavity using segmentation of irregular slopes, and improved empirical equations for the computation of LS factor [23, 24].

So far traditional soil erosion monitoring has been undertaken using field-based sampling methods utilizing discrete spatial intervals. These methods are unable to provide spatially distributed information on land conditions due to the high processing demands and effort involved in analyzing the relevant land properties [25]. However, Remote Sensing and GIS applications are often considered as cost-effective techniques [26] for the collection of data over large areas that would otherwise require a very large input of human and material resources. It can potentially provide spatial products for use in the assessment of soil condition and it has long been recognized [27] as a highly capable method for discriminating soil properties. A field study confirmed that satellite Remote Sensing data can be rapidly processed with computers provides further opportunities for the analysis and interpretation of data, resulting in the acquisition of valuable information over large areas for policy formulation, planning, and management decisions [25]. Moreover, remote sensing offers an important but as of yet underutilized set of tools to manage the transition towards sustainable land usage [28].

Many soil and water conservation efforts have been implemented by the Ethiopian government and charitable organizations in the past decades in northern Ethiopia, but still, soil erosion becomes major problem; and the severity of the problem is increasing from time to time [1, 11]. Evaluating the implemented technologies and land use systems on soil erosion/soil loss effect using modern appropriate tools is paramount important for future soil management issues. This paper estimated the effect of the applied conservation practices and existed land use dynamics on soil loss by the RUSLE model using Remote Sensing and GIS.

2. Materials and method

2.1 Characterization of the study area

Bahir Dar Zuria district is located within 29^0 27′ 34″, 35^0 58′ 40″ East of longitude and 13^0 38′ 19′, 12^0 1′ 37” North of latitude (**Figure 1**). It is about 578 km northwest of Addis Ababa. The district has 32 peasant associations and covers a total area of 128,360.48 ha. It is located in the Amhara region, Ethiopia. It is bounded by Lake Tana, Yilma and Dense Wereda in South, Metcha, and Achefer Wereda in the West and river Abay in the East.

The study area has four major soil types (**Figure 2**). Vertisols, Nitosol, Luvisols, and Cambisols. Vertisols cover 85,394.9 ha (67.7%); Cambisols cover 13,901.5 ha (11%); Nitosols cover 26,313.5 ha (20.8%); and Luvisols cover 496.5 ha (0.5%) of the total area [29]. The distribution of the Vertisols is observed mostly in the plain. Luvisols have a little share when it is compared with the other types of soil. The area around Lake Tana basin is dominated by three geological events: Quaternary Basalts, Oligocene to Miocene basalts, and Quaternary Alluvial and lacustrine deposits. Also, Basaltic lavas of the Aden volcanic series formed the plateau during the Quaternary period of the Cenozoic era. Lake Tana is considered to have been created by the barrier of extended lava of this series.

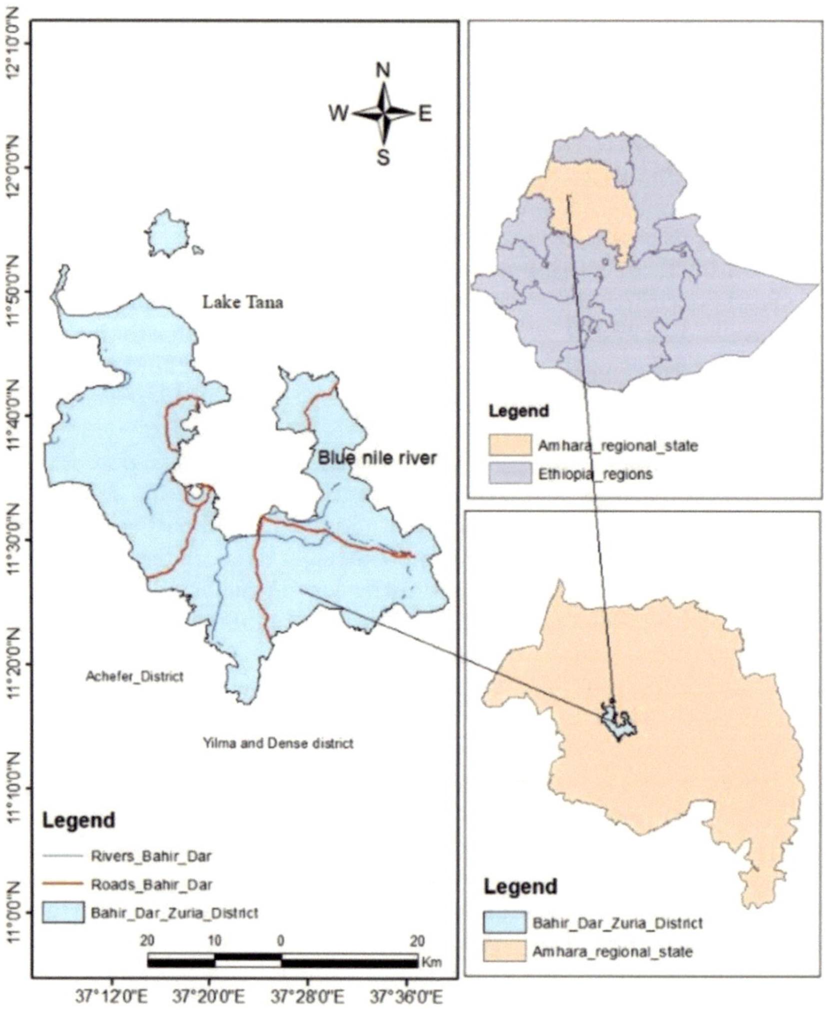

Figure 1.
Map of the study area.

The area lies within the central highland of Ethiopia. About 79% of the total area is found in the slope < 5% and 9.8% is found between 5 and 10%. The remaining area is found in the slope > 10% (**Figure 3**).

The area receives a mean annual rainfall of 1447 mm ranging from a maximum of 2036 mm to a minimum of 895 mm (Amhara region meteorological agency 2018). The study area receives maximum rainfall in summer (June–August). The districts experienced a warm temperature climate, with an average temperature of 21.3 °c. The highest temperature is recorded from February to March and the lowest temperature is observed in January and December (**Figure 4**).

2.2 Data collection and preparation

Using different tools such as Arc GIS 10.3, USA, Erdas Imagine 14.0.0.166, USA, and Garmin 64 handheld GPS, elevation, latitude, longitude, slope, and land use data were collected. To geo-reference the satellite images and digitizes the different features in the image, the topographic map of the study district (1:250,000 scale) was taken from the GIS team of the Amhara region [Amhara region GIS team, 2018].

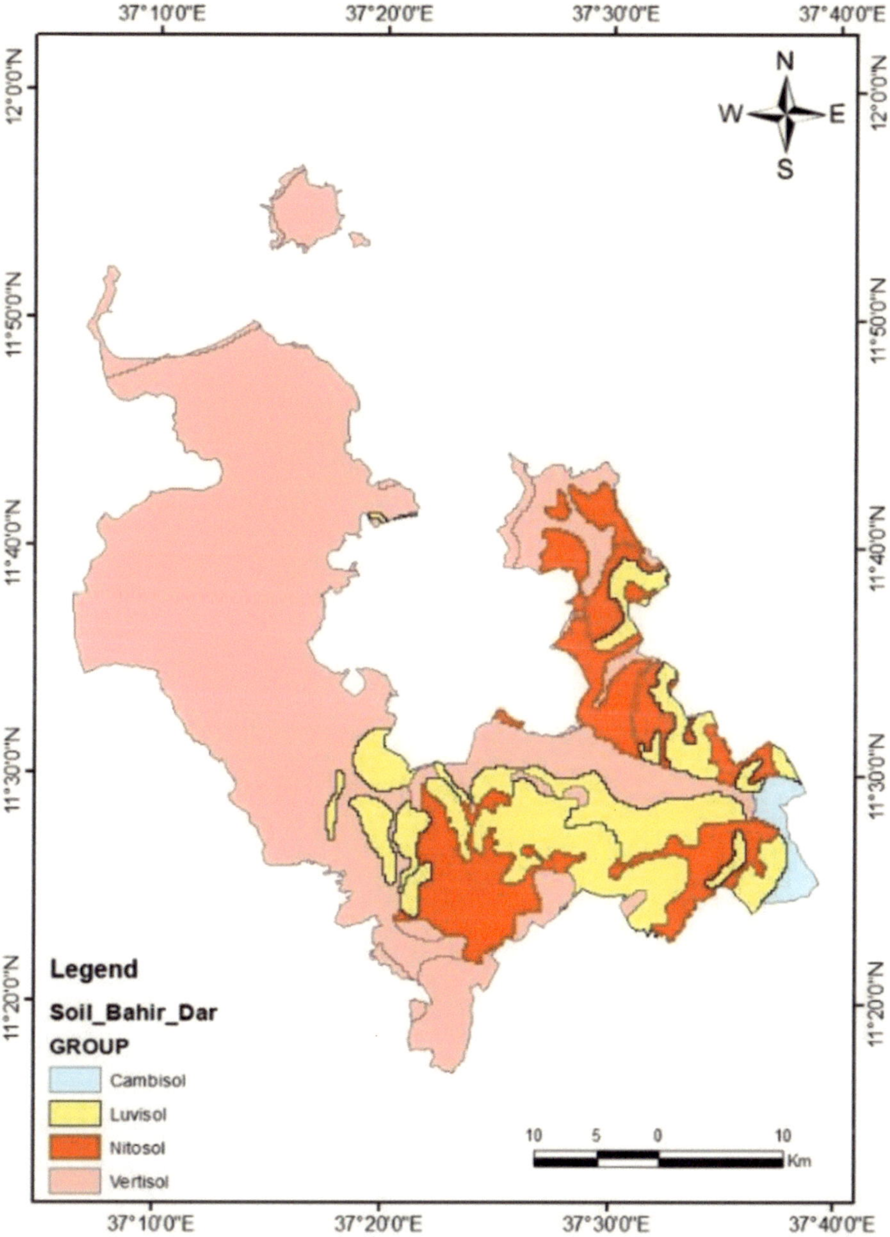

Figure 2.
Soil distribution of Bahir Dar Zuria district.

To calculate the K-factor and produce a soil type map which can be used in the Revised Universal Soil Loss Equation (RUSLE) model, soil map of the study areas was taken from the same sources [Amhara region GIS team, 2018]. The DEM and slopes were generated from the SRTM (Shuttle Radar Topographic Mission) which had been also taken from the same sources [Amhara region GIS team, 2018]. To calculate the RUSLE, R-factor, and to observe the effect of rainfall on soil erosion 19 years of rainfall data of four major rainfall stations w collected from the department of the meteorology of the Amhara region (Amhara region meteorology, 2018). To calculate the C-factor of the RUSLE model satellite images of 2018 had been downloaded from the internet (date accessed:13/02/2018). To calculate the LS factor of the RUSLE

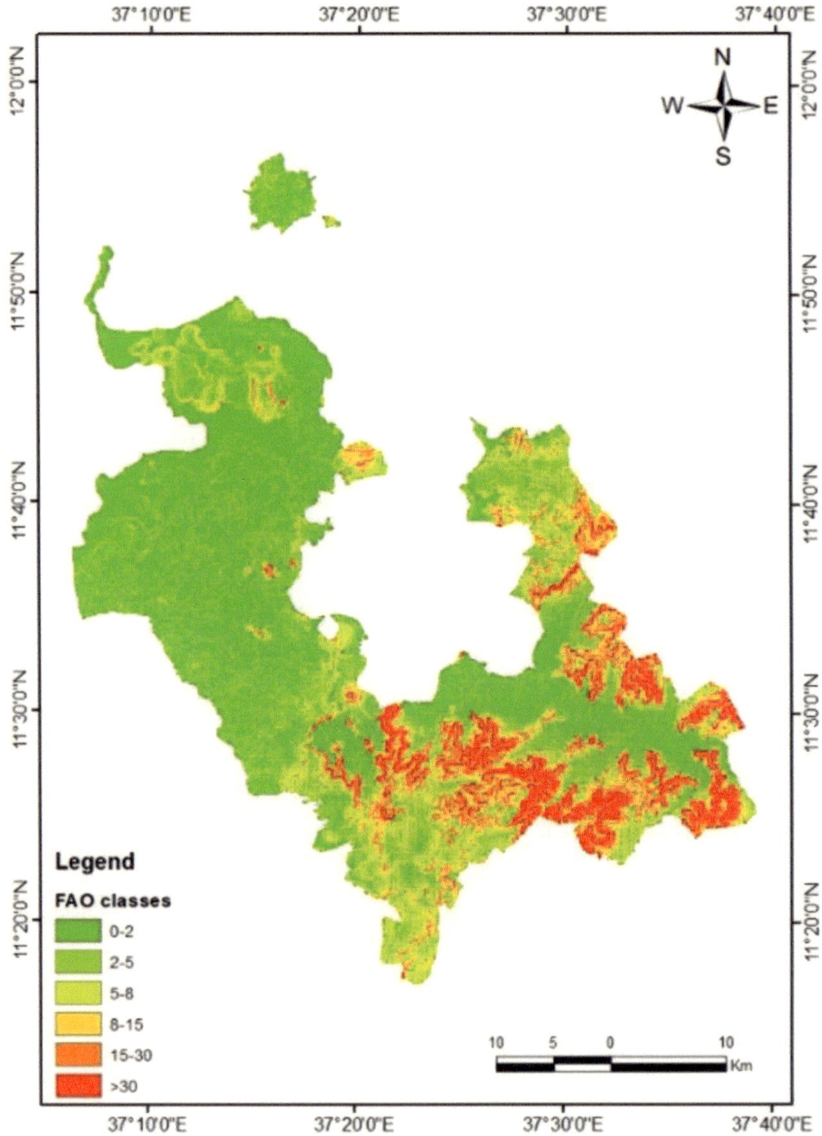

Figure 3.
Slope distribution of Bahir Dar Zuria districts.

model, Digital Elevation Model (DEM) was converted from the SRTM resolution of 30 m data.

Field surveys on different land use/land cover dynamics, slope steepness, and soil erosion conservation practices were made. Moreover, field observation of vegetation and biodiversity under certain land use/management practices and ground points were taken.

The RUSLE is a model that can predict the long-term average annual rate of soil erosion on a field slope as a result of rainfall patterns, soil type, topography, crop system, and management practices [30]. The strength of this model permits an all-inclusive analysis by breaking down soil erosion into elements. Rainfall, soil, slope, land use/land cover data, and management practice were collected and hence, used for the estimation of soil loss using the formula.

The equation is presented as

$$\mathbf{A} = \mathbf{R} * \mathbf{K} * \mathbf{LS} * \mathbf{C} * \mathbf{P}, \tag{1}$$

Where: **A**- represents soil loss in tons/ha/yr., **R**- Rainfall erosivity is a term used to describe the degree of soil loss due to rainfall effect when other factors of erosion are held constant. It is an index that characterizes the effect of raindrop impact and the rate of runoff associated with a rainstorm. The erosivity index, R, depends upon the amount and intensity of rainfall. It is very high where frequent heavy storms occur and declines as the amount of rainfall and intensity of storms diminish. R is calculated from long-term rainfall data. A high correlation r = 0.88 for monthly precipitation and monthly erosivity was found together with the following regression equation:

$$R = -8.12 + 0.562 * P \tag{2}$$

Where P is the mean monthly precipitation in millimeter.

K- The K-factor is defined as the rate of soil loss per unit of R-factor on a unit plot [31]. The K-factor also defines as the resistance of the soil to both detachment and transport, the unit depending upon the amount of soil occurring per unit of erosivity and under specified conditions. The inherent properties of the soil would have more influence on being liable to erosion than other factors. RUSLE K-factor depends on a combination of soil and climatic parameters developed under specific conditions in the USA, which might not be suitable to different conditions in other parts of the world, such as in the Ethiopian condition. For the Ethiopian case according to Hurni [8], the determination of the K-factor was simplified by giving the soil color representing a major soil type a specific value. Hence to calculate the K factor soil data was obtained from woody biomass and the value for different soil was given according to Hurni [8] adaptation to the Ethiopian condition. The spatial variation of the K-factor was determined using the soil maps produced by the Woody Biomass [29]: using GIS attribute table level editing which was adapted to Ethiopian conditions by Hurni [8]. The resulting shapefile was changed to a grid file using convert feature to raster.

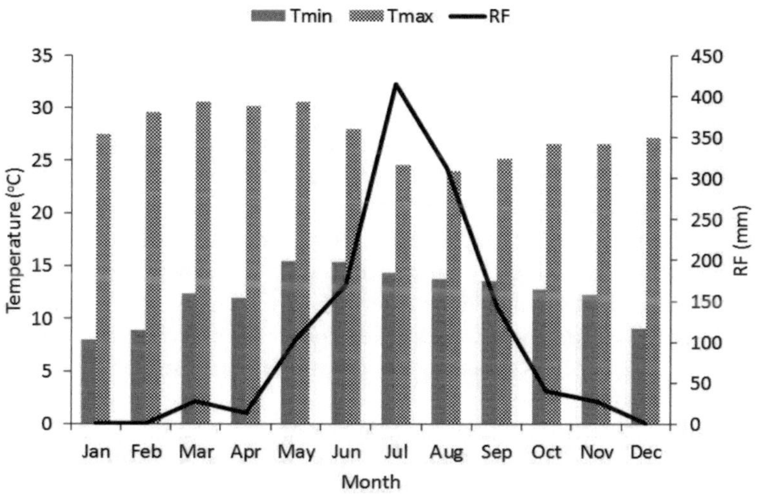

Figure 4.
Monthly rainfall, minimum and maximum temperature of the study area.

L & S - the topographic factor, is divided into 2 components: S is the slope grade expressed as a percentage and L is the length of the slope. Slope grade affects mainly the speed of runoff. Slope length affects mainly the amount of runoff. In assessing the effects of slope length it is necessary to take into account the total length of the slope over which runoff occurs, not just the length of the field in question. Hurni [8] calculated the L and S factor depending on its slope length. For the calculation of the L-factor, there should be slope data. The slope length and steepness (LS-factor) were derived from the DEM of 90-meter resolution. The DEM was converted into fill sinks and flow direction grid using a hydrological extension of Arc GIS version 10.3. Secondly, a flow accumulation grid was created using the flow direction grid in the same technique. The third step was to calculate the LS- factor using the flow-accumulation grid and the slope grid using the same method. Generally, the DEM was used to generate slope, fill sinks, flow direction, flow accumulation, and LS maps using the Arc Hydro extension.

C- Cropping practices have a strong influence on erosion by their effect on the amount of protective coverage that crops and crop residues provide. The C-factor is defined as the ratio of soil loss from land with specific vegetation to the corresponding soil loss from continuous fallow [30]. The RUSLE C-factor is a measure of the cropping and management practices' effect on soil erosion [31]. The more of the soil that is left uncovered the greatest the risk of soil erosion by either wind or water and vice versa. To calculate the C- factor land use/ land cover for image 2018 was collected and the resulting data was substitute with the value given by Hurni [8] for different land cover types. Hence the C-value for vegetation is considered as C = 0.02 and for grazing land is C = 0.03 and for degraded land is C = 0.2 and for cropland is C = 0.16.

P- is the support practice factor. It reflects the impact of support practices on the average annual erosion rate. It indicates the fractional amount of erosion that occurs when any special practices are used compared with what would occur without them. The P-factor gives the ratio between the soil loss expected for a certain soil conservation practice to that with up-and down-slope plowing [30]. Special conservation practices have the effect of reducing erosion. The support practice factor is the ratio of soil loss with a support practice like contouring, strip cropping, or terracing to soil loss with straight–row farming up and down the slope. Hence the different support practices methods that are observed in the study area were collected and their values are substituted with the value given by Hurni [8] for different management practices in the adaptation of RUSLE to the Ethiopian condition.

2.3 Methodology

To assess and analyze the effect of different variables on a single area, the weighted overlay method was conducted on the RUSLE variables (**Figure 5**).

Digital Image processing

Digital image processing involves numerous procedures including formatting and correcting of the data, digital enhancement to facilitate better visual interpretation, had been carried out [32]. The common image processing functions available in image analyses like radiometric correction, geometric correction, image mosaic, subsetting, and image enhancement had been made accordingly.

Image Classification

Image classification is defined as the process of sorting pixels into a finite number of individual classes, or categories of data, based on their data file values [32]. If a pixel satisfies a certain set of criteria, the pixel is assigned to the class that corresponds to those criteria. This process is also referred to as image segmentation. Depending on the type of information you want to extract from the original data, classes may be associated with known features on the ground or may simply represent areas that look different to the computer. An example of a classified image is a land cover map, showing vegetation, bare land, pasture, etc. [33].

To classify the images, ground truth points were collected from different land use land cover types, and hence supervised image classification technique was applied. Supervised training is closely controlled by the analyst. The sample ground truth points taken during the field time help a lot to identify the land use/land cover on the image by using supervised classification and hence the computer can automatically classify the image based on the given sample ground truth points.

Rainfall Erosivity Factor

The erosivity factor R was calculated according to the equation given by Hurni [8], for Ethiopian conditions based on the easily available mean annual rainfall (P). $R = -8.12 + 0.562^{*}P$; Where P is mean annual rainfall in mm. The correspondence R values of the four stations were calculated as follows in **Table 1**.

The above data with their location were used to generate a rainfall erosivity map using Arc GIS 10.3 Spatial Analyst, IDW Interpolation. The R-map was simply generated from the mean annual rainfall data.

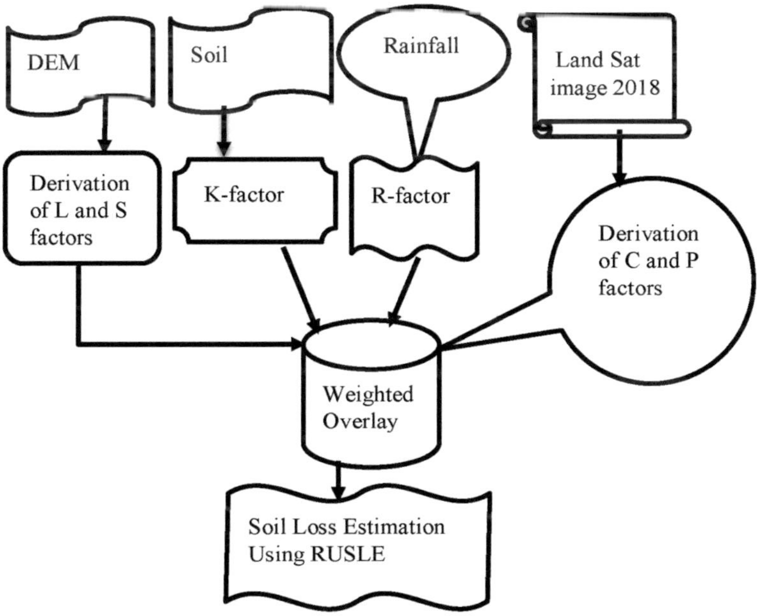

Figure 5.
Methodology flow chart of soil loss estimation using RUSLE. Variables like soil, rainfall, and slope, which have a great relationship with land degradation were analyzed at the same time to assess land degradation.

Station's name	Mean annual rainfall	R-value
Bahir Dar	1503.8	837.1
Meshenti	1280.5	711.5
Tis Abay	1968.3	1098.1
Zege	1470.5	818.3

Table 1.
RUSLE, R-factor.

Soil Erodibility Factor (K)

The K-factor is defined as the rate of soil loss per unit of R-factor on a unit plot [31]. Hurni [8] in the adaptation of RUSLE to Ethiopian conditions considered the soil color to calculate the K-value. The resultant soil color and their K-value are presented in **Table 2**.

The spatial variation of the K-factor was determined using the soil maps produced by the Woody Biomass [29]: using GIS attribute table level editing which was adapted to Ethiopian conditions by Hurni [8].

Topographic (LS) Factor

The analyzed LS factor with the slope ranges is presented in **Table 3**. After generating the flow accumulation the topographic factor (LS) factor in the GIS environment was used [30].

$$LS = (\text{Flow accumulation} * \text{Cell size} / 22.13) \wedge 0.4 * (\sin(\text{slope} / 0.0896) \wedge 1.3 \tag{3}$$

Where flow accumulation refers to the number of cells contributing to flow into a given cell and cell size is the size of the cells being used in the grid-based representation of the landscape [30, 34] (**Figure 6**).

Crop Management (C) Factor

The C-factor was given based on the estimated value that was developed by [8]. The mean value of different crops (C = 0.16) had been taken for the C - value for croplands. Besides, the C-value for vegetation was considered as C = 0.02 and for grazing land was C = 0.03 and for degraded land was C = 0.2. Thus, the C-value was applied to the land use map of the 2018 land sat image. After generating the classified land sat image of 2018, the format was changed into a vector and a corresponding C-value was assigned to each land use type using the editing menu of Arc GIS 10.3 from the C-value adopted by Hurni [8] for the RUSLE model.

Conservation Support Practice (P) Factor

The data related to management practices were collected during the fieldwork. Values for this factor were assigned considering local management practices and

Soil color	K-value	Soil type
Black	0.15	Vertisols
Brown	0.20	Luvisols and Cambisols
Red	0.25	Nitosols

Source: [8].

Table 2.
K-values of soil colors in the study areas.

Slope class %	0–5	5–10	10–20	20–30	>30
Area(ha)	102,242	12,544.5	9062.3	3350.2	1161.5
L-factor	0.4	0.7	1.0	1.2	1.4
S-factor	0.4	1.0	1.9	3.0	3.8
LS-factors	0.16	0.7	1.9	3.6	5.32

Table 3.
RUSLE LS factor.

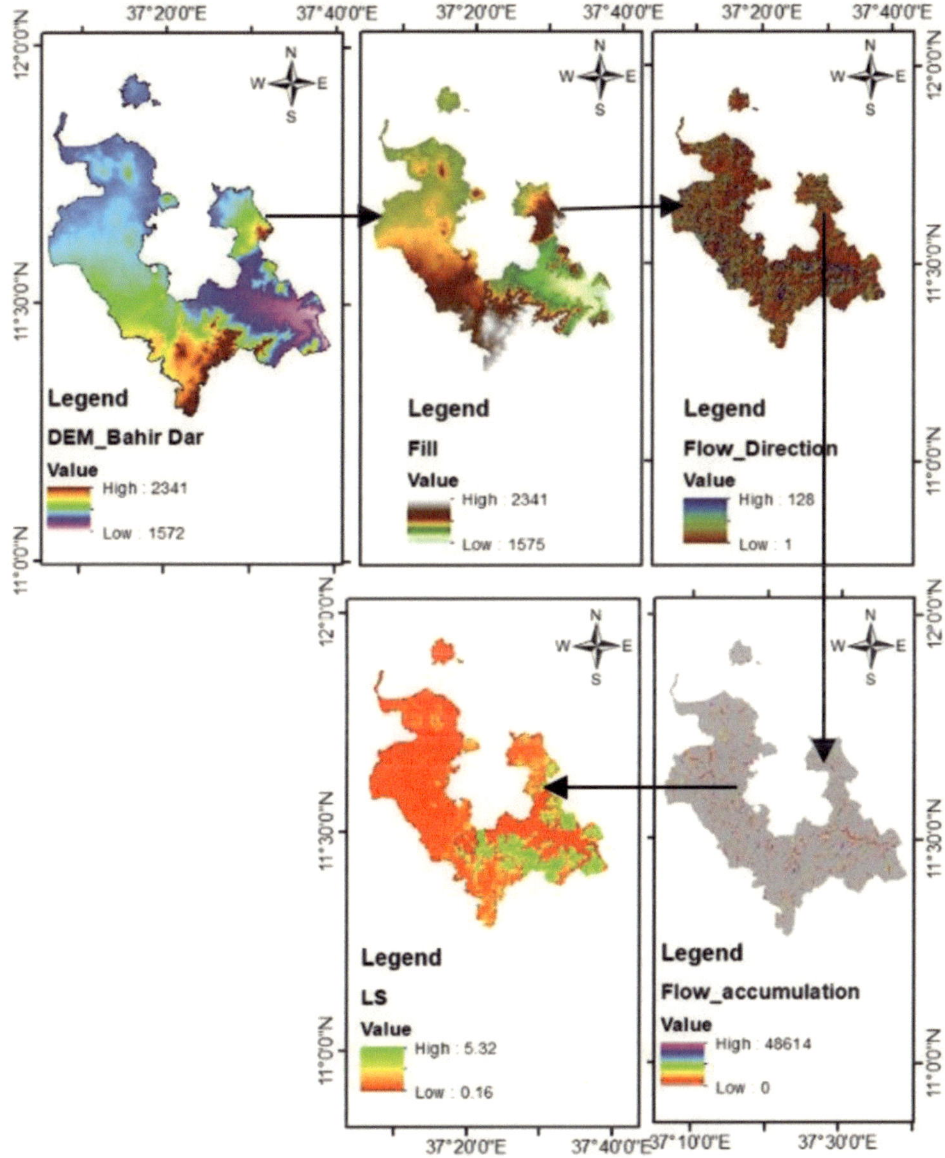

Figure 6.
Flow chart of LS factor.

based on values suggested in Hurni [8]. The result land cover map and the associated P-factors were used to generate a grid surface for the P-factor, utilizing Arc GIS 10.3 spatial analyst. Soil bund, bench terrace, grassed waterways, grassed strips, and area closure are the dominant management practices. The bench terraces and area closures are mainly found at the higher elevation, whereas soil bund and grassed waterways are distributed in the low laying areas.

Overlay

The RUSLE is an index method that includes factors that represent how land cover, climate, soil, topography, and land use affect soil erosion caused by raindrop impact and surface runoff. To assess the effect of these parameters at the same time, the overlay of all factors in a single scene was made. This was done utilizing Arc GIS software (**Figure 7**).

3. Results and discussion

3.1 Revised universal soil loss estimation (RUSLE)

3.1.1 Rainfall erosivity (R-factor)

The analysis of the monthly average rainfall erosivity revealed that more rainfall occurs during summer (**Figure** 7). During this period, the height of agricultural production is small, which leaves the soil surface unprotected against raindrop impacts, resulting in a high risk of erosion in areas of cultivation. This indicates that when other erosion parameters are held constant, areas with the highest rainfall have high R-values (**Table 1**) and are exposed to erosion. As it is presented in **Figure 8**, the Northern and Eastern parts are more exposed to erosion than many other areas in the study area when other parameters remaining constant. In principle the greater the R-value the greater the soil loss is and the opposite is also true. When other soil loss factors are remaining constant, greater soil loss is observed in areas where high R-values were registered.

Using the R-value, rainfall distribution throughout the study area was interpolated. This method was designed in a GIS environment with the principle of things found to be close to one another are more alike than those that are farther apart. Different findings reported that rainfall erosivity has been one of the leading factors for soil erosion. Among them, Outhman *et al.* [25] concluded that rainfall erosivity increase soil erosion especially when the heights of the agricultural lands are short. Also [6, 7, 35] reported that rainfall erosivity has a great role in aggravating soil erosion and soil loss.

3.1.2 Soil erodibility (K-factor)

The K-value for Vertisols, Luvisols, and Cambisols and Nitosols are 0.15, 0.2, and 0.25 respectively (**Figure 9**). This is to means that as the K-value increases the erodibility of the soil also increases and vice versa. In this case, the Nitosols for example are more erodible than vertisols. In the same way areas with Nitosols are more vulnerable to erosion than areas with Vertisols. Vertisols, Nitosols, Cambisols, and Luvisols had the first, the second, and the third share in the study areas respectively. The Nitosols in the study area is observed in the higher slopes i.e. slope > 10% and covers 10.6% of the study area, whereas the Vertisols are observed in lower slopes i.e. slope < 5% and covers 79.6 percent. The remaining part of the study area has been covered by Nitosols and Vertisols. Soil type has a great role in soil erosion because some soils are more erodible than others. In this case, considering the soil erodibility factor in RUSLE parameters helps to see its effect on soil erosion.

Different researchers noted that soil erodibility is one of the leading factors to soil erosion. For instance, Assen [36] reported that Nitosols and Cambisols are more erodible to soil erosion than other soil types. Hurni [8] also founded similar findings.

3.1.3 Topographic (LS) factor

The LS-factor value represents the relative erodibility of the particular slope length and steepness (**Figure 10**). The LS factors have a great impact on erosion. Higher slopes have higher LS value and lower slopes have lower LS value. In the same way, high LS values indicate that higher soil erosion and the opposite being

DOI: http://dx.doi.org/10.5772/intechopen.95393

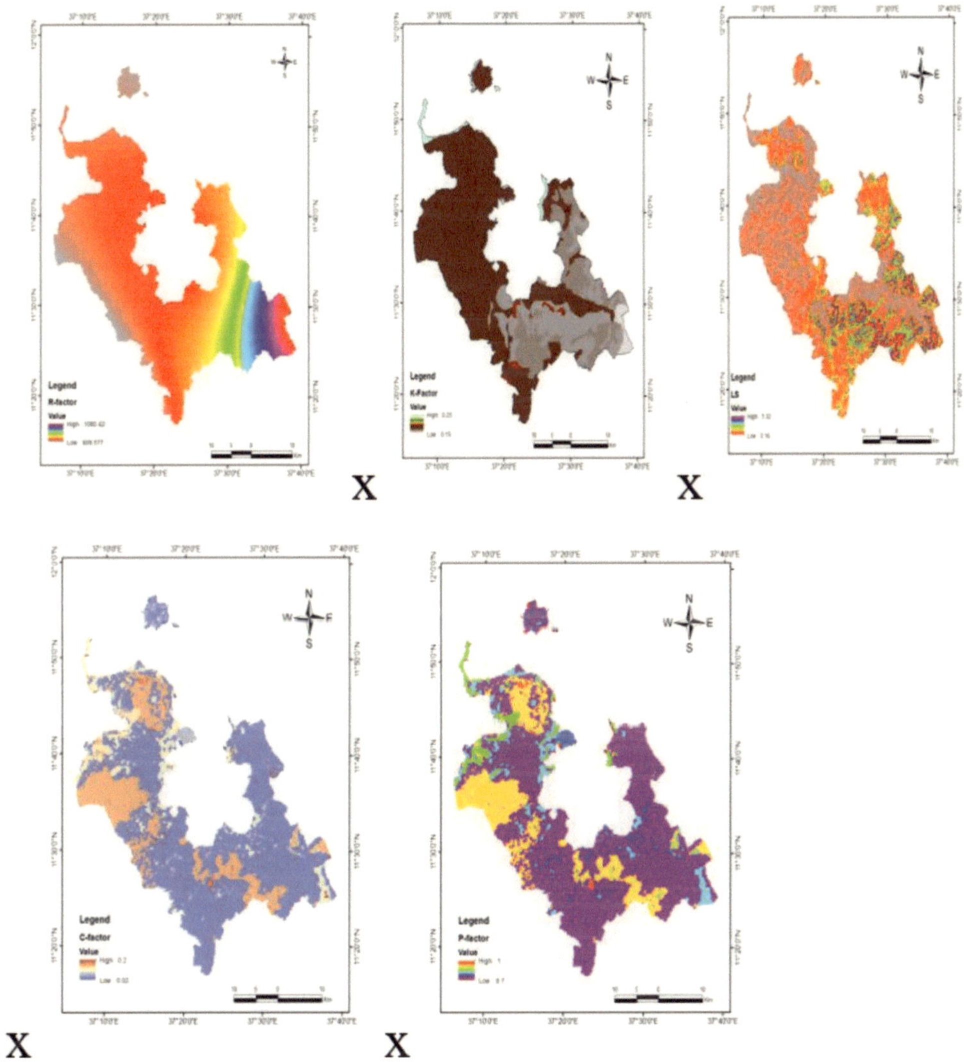

Figure 7.
Overlay of RUSLE parameters.

other factors of erosion remaining constant. As it is presented in **Figure 10**, a higher LS factor value (LS = 5.32) is observed in the Central and South margin of the study area. On the contrary, a lower LS factor value (LS = 0.16) was observed in the plains of the North and northwest part of the study area. Therefore, other factors remaining constant in the Central and -south margins of the study (where LS-factor is greater) area are at high risk of soil erosion than any other area. Various findings confirmed that the LS factor have a great impact than any other RUSLE parameters. For instance, Outhman *et al*. [25] Palestine reported that the LS factors are the two major factors of soil erosion than any other factors.

Yitaferu [35] also presented the effect of the slope in soil erosion separately from the other parameters [37]. This indicates that the slope has a great impact on soil erosion than any other parameter. Besides, the report by [6, 7, 38] showed that soil erosion increase as the slope increases. Generally, as the slope increase, the soil loss also increases unless a special soil loss conservation mechanism is applied in the higher slopes.

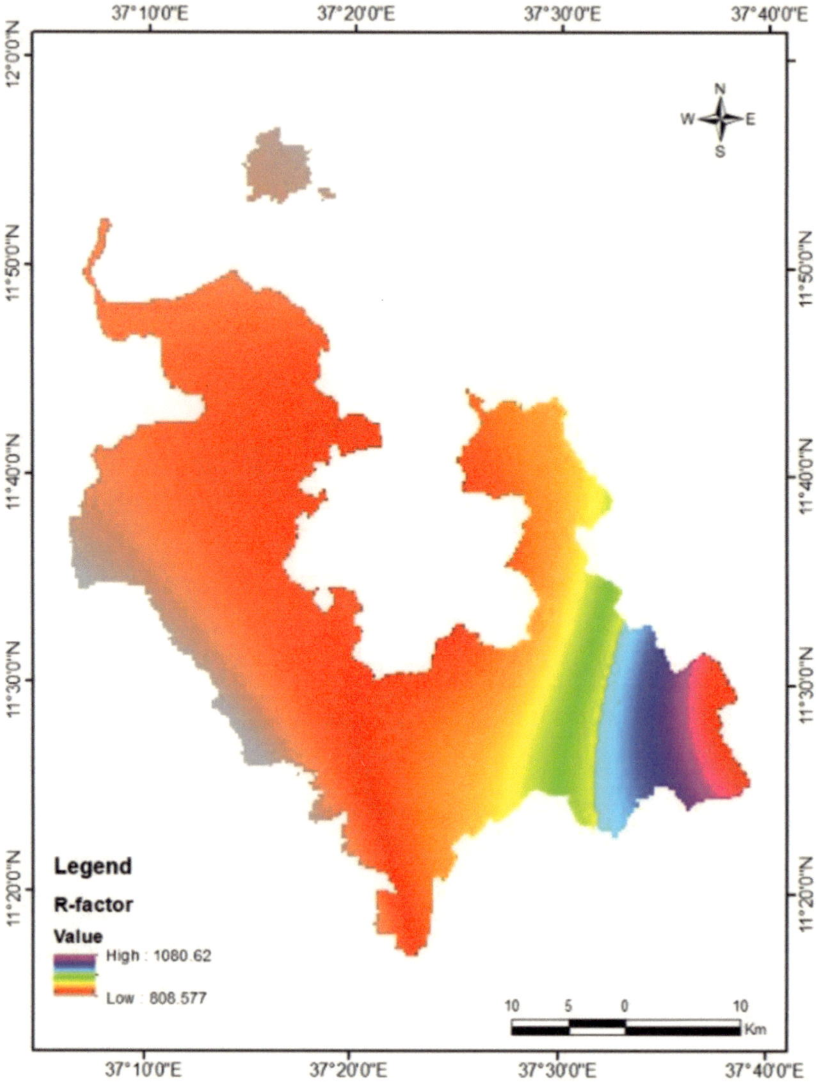

Figure 8.
Rainfall erosivity map.

3.1.4 Land cover (C-factor)

The C-values for agriculture, vegetation, grassland, and degraded land were 0.16, 0.025, 0.03, and 0.2, respectively (**Figure 11**). This is to means that as the value for the C-factor increases the capacity of the area to resist erosion decreases and vice versa. For example, the degraded lands are more vulnerable to erosion than vegetation areas because the C-value for degraded land and vegetation are 0.2 and 0.025, respectively. It is true that soil erosion increase, if there is no cover or if the cover is not resistant to erosion. For instance, [9] reported that differences in the vegetative cover have been mainly responsible for the variation in erosion rates in the Ethiopian highlands. Morgan [39] also reported the differences in erosion rates caused by different land use practices on the same soil are much greater than the corresponding changes from different soils under the same land use. From this

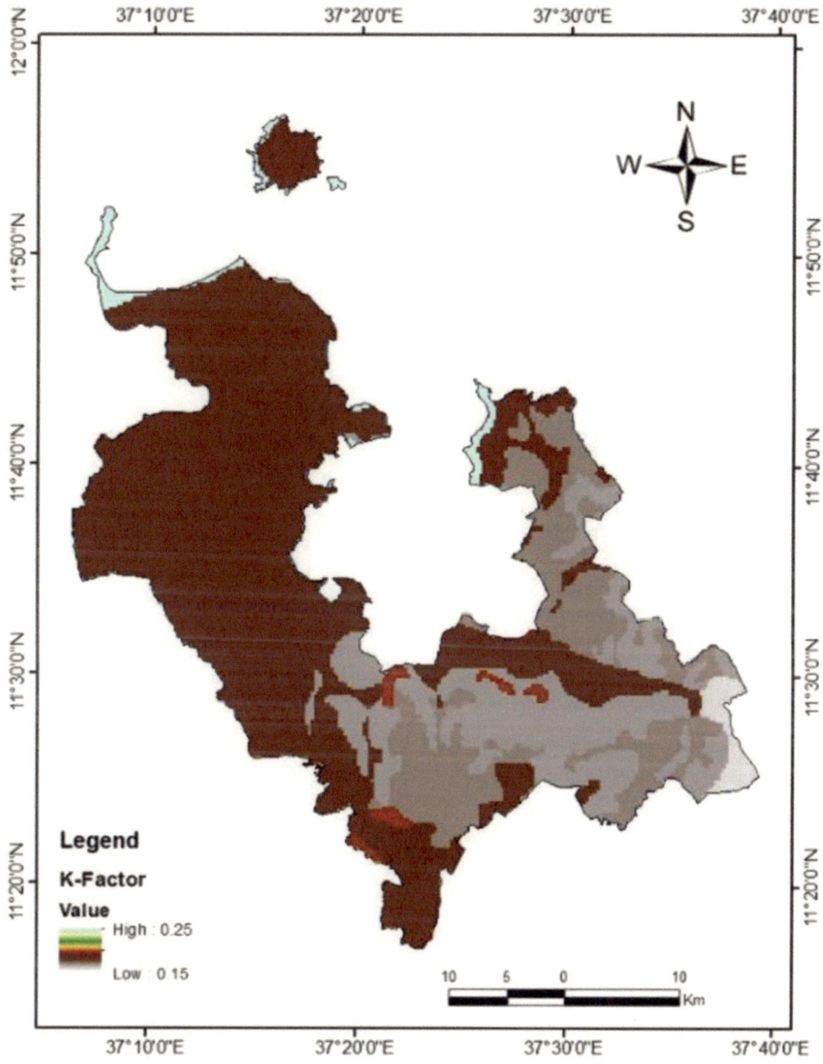

Figure 9.
Soil erodibility factor map.

study, we can understand that areas having a higher value of C-factor have a higher capacity for soil loss resistance. This is to means that areas with vegetations or any other cover types have less soil loss than areas with barren land.

3.1.5 Conservation support practice (P) factor

The support practice factor (P) reflects the impact of support practices on the average annual soil loss rate. It indicates the amount of soil loss that occurs when any special practices are used compared with what would occur without management. According to our study, the P-value for agriculture, bareland, vegetation, and grasslands were 0.9, 0.7, 0.8, and 1, respectively (**Figure 12**). This is to means that areas having conservation practice have the lowest erosion than areas with no conservation practice because, in areas where there is conservation practice, runoff speed could be reduced.

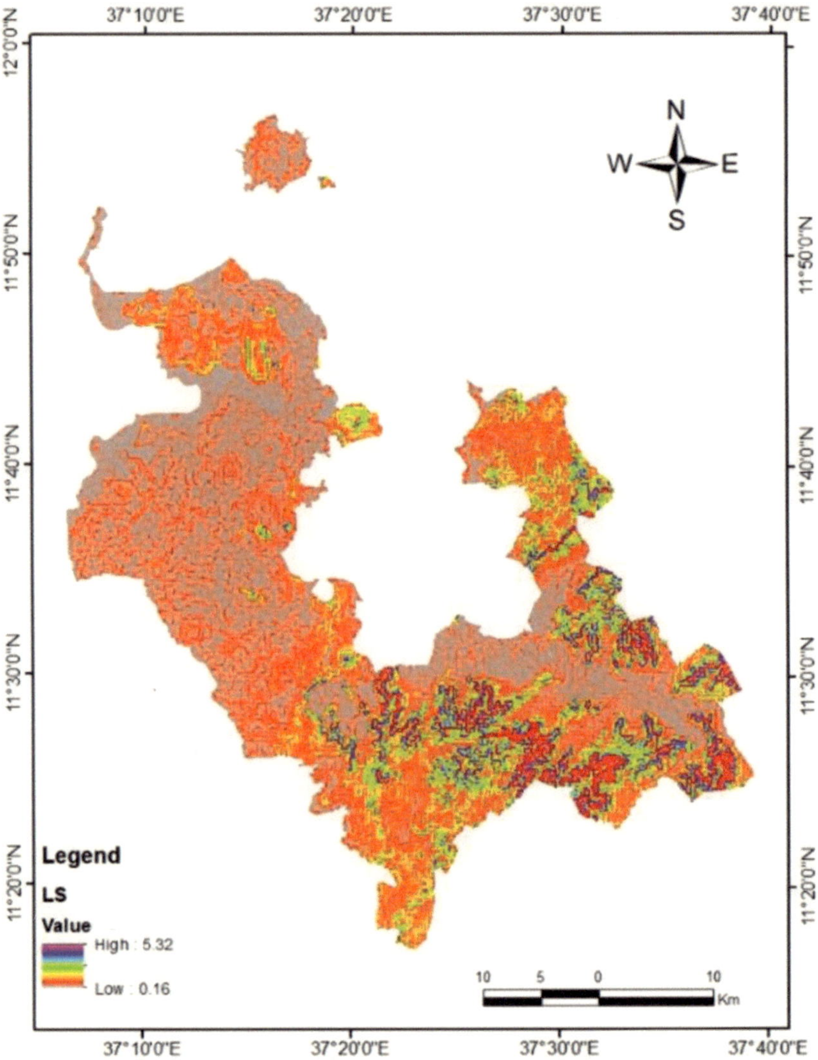

Figure 10.
Topographic factor map.

To reduce soil erosion, conservation practices have been implemented in the study areas. Many researchers confirmed that conservation practice have a significant role to reduce soil erosion [6, 11, 25, 35].

3.1.6 Annual soil loss

To ease the presentation of the output data, the result considers three main categories such as annual soil loss based on slope gradient, annual soil loss based on land use/land cover type, and annual soil loss based on slope and land cover.

3.1.6.1 Annual soil loss based on slope gradient

The slope has a major role in the RUSLE model since it determines the direction and velocity of the water movement. It also determines the processes of detachment, transport, and accumulation of soil particles. We have found that as the

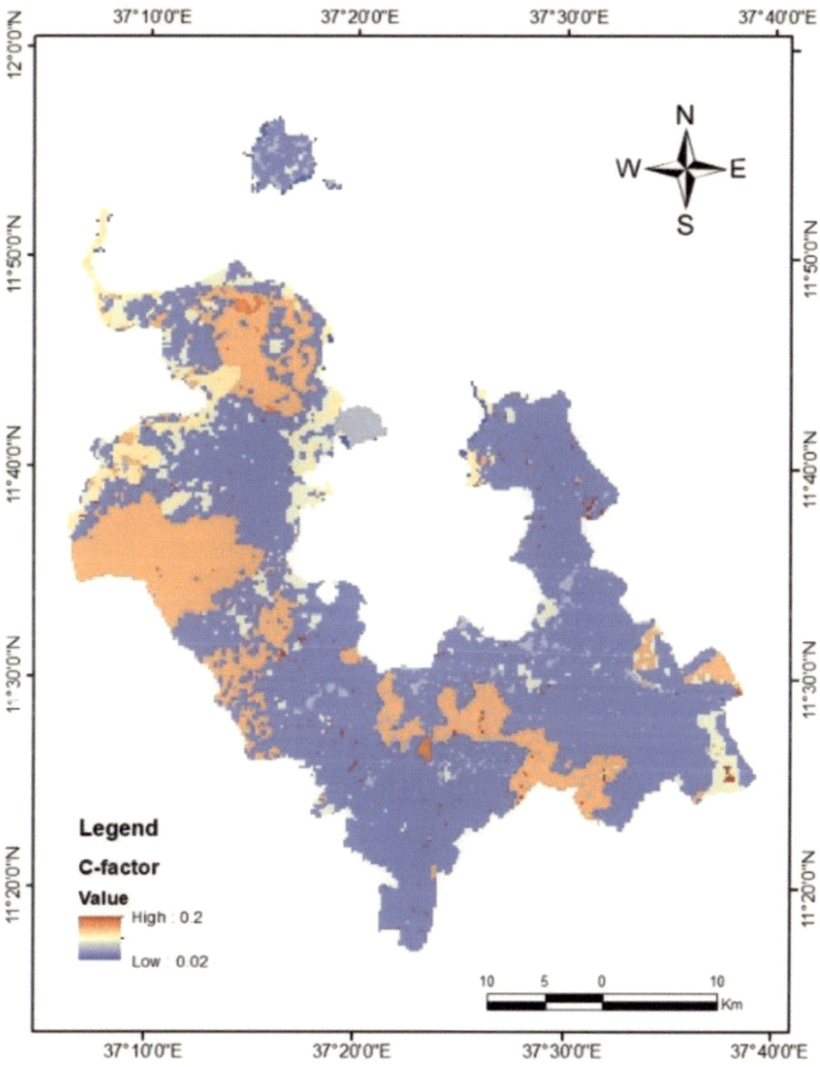

Figure 11.
Cropping practice map.

slope increases the amount of soil loss also increases (**Table 4**, **Figure 13**). This is because higher slopes increase the speed of water and transport of soil particles. Slope and soil losses have a direct relationship i.e. as the slope increases the annual soil loss also increases. A relatively small amount of soil loss per hectare of land was recorded around the low slope areas whereas a high amount of soil loss per hectare had been obtained at sloppy lands. As it is observed in **Figure 13**, 79.65% of the study area experiences low soil loss which is <1.2 ton/ha/yr. This indicated that most of the area is situated in the plains and have low soil loss.

Conversely, soil loss is very high for slopes >30%. This indicates that slope has a great impact on regulating soil loss.

3.1.6.2 Annual soil loss based on land cover type

The type of land cover has a great impact on soil loss estimation and various scientists tried to relate the RUSLE soil loss estimation model with the land use

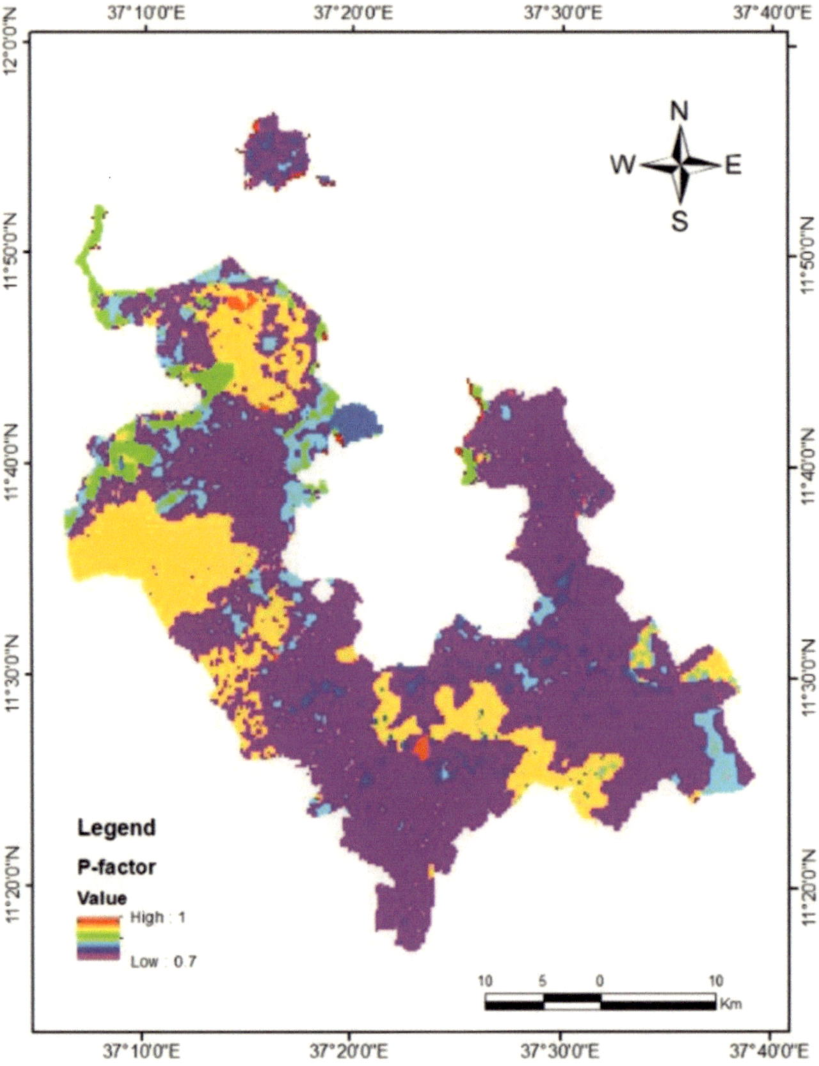

Figure 12.
Conservation support practice factor map.

Slope %	Area (ha)	%	Soil loss in ton/ha/year
0–5	102,242	79.7	< 1.2
5–10	12,544.5	9.8	1.2–5.2
10–20	9062.3	7.1	5.2–6
20–30	3350.2	2.6	56–12
>30	1161.5	0.9	117–192

Table 4.
Soil loss estimation based on the slope.

dynamics. As is presented in **Table 5** and **Figure 14**, the annual soil loss for cropland, vegetation, grassland, and degraded land was 19.05, 8.78, 8.82, and 71.16 ton/ha/yr., respectively. This is to means that, the type of land cover have great relationships with the amount of soil loss. For example, the soil loss under cropland

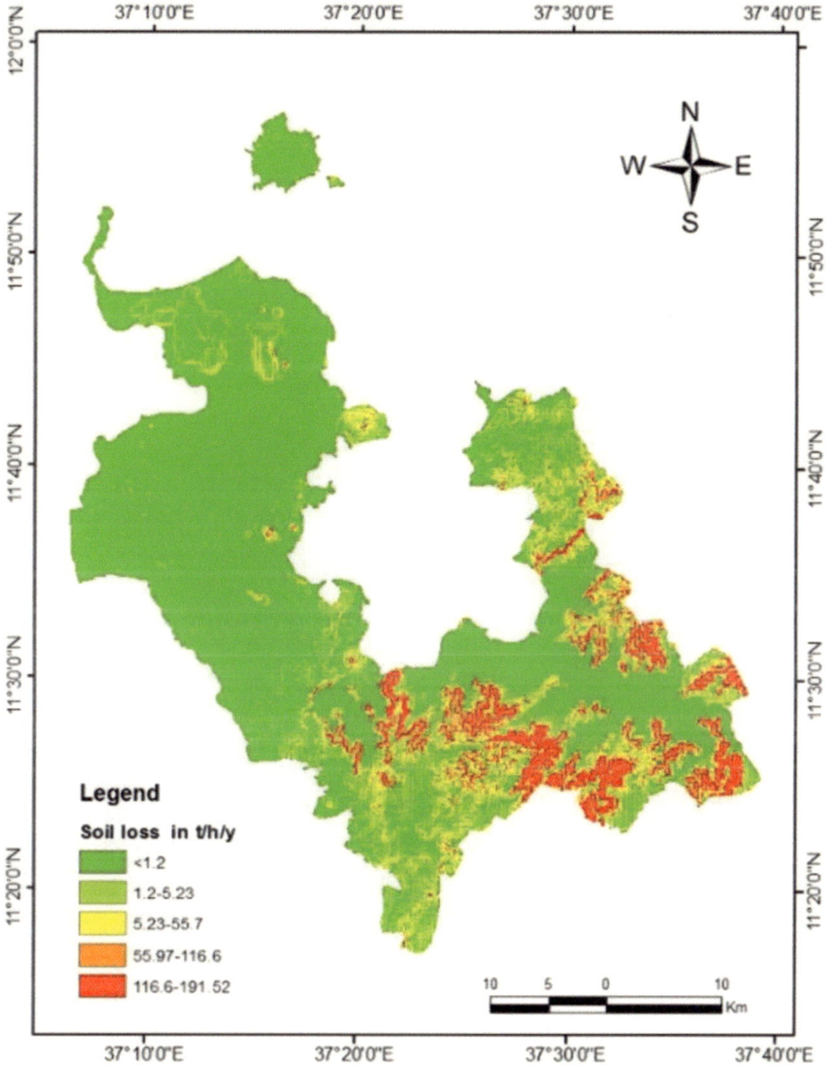

Figure 13.
Soil loss based on slope gradient.

Land use/land cover	Area (ha)	%	Annual soil loss in t/h/y
Crop land	68,218.6	53.15	19.05
Vegetation	11,861.2	9.24	8.78
Grass land	29,774.3	23.2	8.82
Degraded	17,315.4	13.49	71.16

Table 5.
Annual soil loss estimation for different land cover.

was more than the soil loss for vegetation and this means that areas covered with vegetations have less vulnerable to erosion than areas covered with crops.

Similarly, the soil loss for degraded land was greater than the grasslands, vegetation, and crops. On the contrary, vegetation cover and grasslands were more erosion resistant than croplands and degraded land.

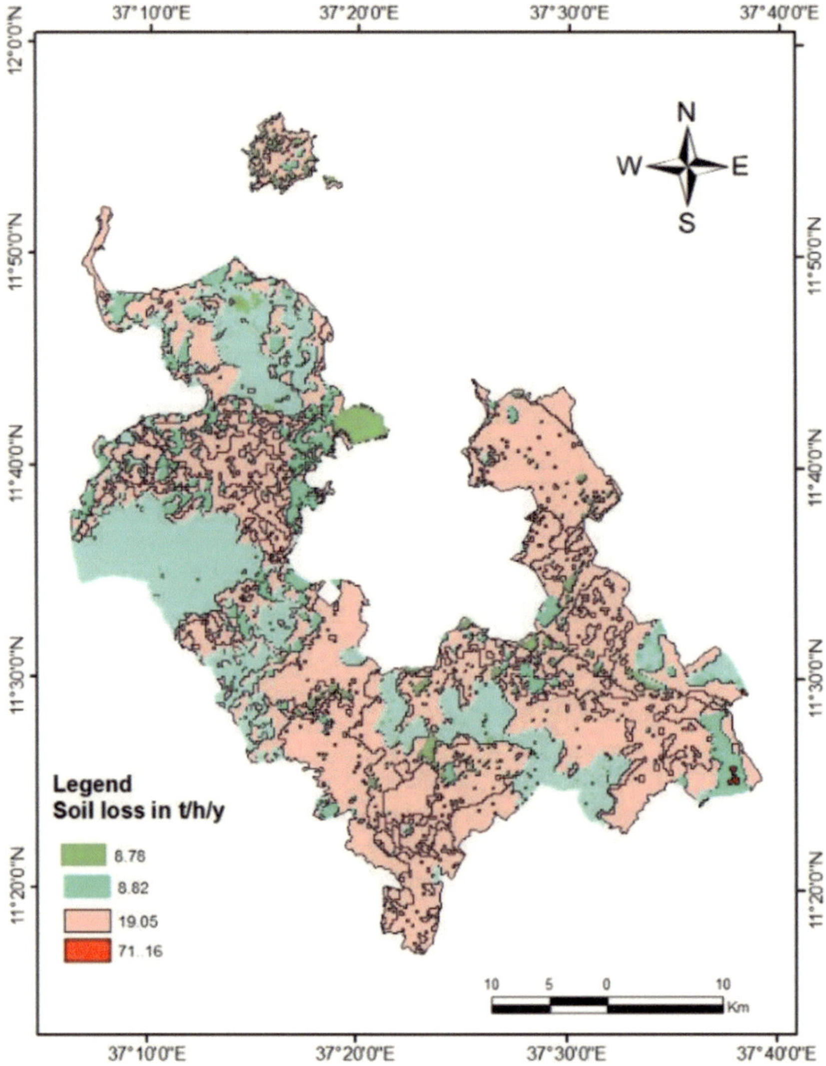

Figure 14.
Soil loss of some land cover types.

Our result agreed with the finding of Hurni [40] who studied the effect of different land use/land cover types on soil loss in Ethiopia. According to his report, the soil loss for cropland, grassland, totally degraded land, and bushland was 42, 5, 70, and 5 ton/ha/yr., respectively. Assen [36] reported that severely deforested and cultivated lands are more vulnerable to erosion, 18 ha of land was exposed to soil erosion every year and 95% of the gullies were also observed in cultivated land confirming the susceptibility of the area to water erosion in general.

Hurni [8] also reported that in Ethiopia cultivated land followed by severely deforested landform the major source of soil erosion. Moreover, Hurni [9] noted that differences in vegetation cover have been mainly responsible for the variation in erosion rates in the Ethiopian highlands. Morgan [39] reported that the differences in erosion rates caused by different land use practices on the same soil are much greater than the corresponding changes from different soils under the same land use.

Slope	Cropland		Vegetation		Grassland		Degraded land	
	Area (ha)	Soil loss in t/ha/y	Area (ha)	Soil loss in t/h/y	Area (ha)	Soil loss in t/ ha/y	Area (ha)	Soil loss in t/ ha/y
0–5	59373.2	<0.12	8255.1	<0.01	21526.9	<0.08	12605.7	<2.35
5–10	4999.9	0.12–0.8	1144.6	0.01–0.15	4198.2	0.08–0.5	2268.4	2.35–10
10–20	2460.7	0.8–4.97	1340.7	0.15–0.7	3513.4	0.5–1.6	1714.3	10–38
20–30	1090.3	4.97–13.77	761.5	0.7–2.6	119.1	1.6–5.2	588.7	38–111
>30	294.7	13.77–35.91	359.3	2.6–6.7	416.8	5.2–11	138.6	111–219

Table 6.
Average annual soil loss based on slope and land cover.

3.1.6.3 Annual soil loss based on slope and land cover

The soil loss with the slope gradient can simply explain the effect of slope in soil erosion by taking the average value of other factors; even though, different land covers at different slope have a great impact on soil erosion [38]. In this study different land covers situated on different slopes with their relative area have been analyzed using Arc GIS 10.

As it is presented in **Table 6**, the same land cover but different slopes have different soil loss amount. For example, the amount of soil loss for cropland in different slopes (0–5 and > 30) varies between 0.117 and 35.91 ton/ha/yr., respectively. This indicated that land cover type can be greatly determined by slope difference.

In an experiment conducted in 2005 and 2006 cropping seasons in northern Ethiopia, a significant difference ($p < 0.05$) in soil loss in wheat and *tef* cropped field was observed [41]. The soil loss reduction at wheat crop was 76% in permanent bed (PB) while 61% in Terwah (TERW) as compared to traditional tillage (TT). Therefore, land cover and slopes can determine the amount of soil loss in a particular area.

4. Conclusion and recommendations

This study assessed soil loss using a GIS-based RUSEL equation. The GIS-baes RUSEL equation well estimated the amount of soil loss in our study areas, which resulted in comparable findings with other findings. The annual soil loss increased at LS and S factors compared with the other RUSEL factors. Compared with other land uses, barelands and croplands that found at the higher elevations generated more soil loss. It is found that lack of vegetative cover during the critical period of rainfall, expansion of croplands, and lack of support practices also increase soil erosion. The application of soil bund, area closure, contour tillage, terraces, and grass strip barriers are suggested to break the slope length into shorter distances, reducing overland flow velocity and soil erosion.

Acknowledgements

We wish to express our gratitude to the Amhara National regional state Meteorological Services Agency and GIS team of the Amhara National regional state for their support of climate data and topographic maps, satellite image data, respectively.

Author details

Nurhussen Ahmed Mohammed[1] and Desale Kidane Asmamaw[2,3]*

1 Wollo University, Ethiopia

2 Department of Natural Resource Management, Bahir Dar University, Ethiopia

3 Department of Environment, Ghent University, Soil Physics Research Group, Belgium

*Address all correspondence to: kidanedesu@gmail.com; desaleKidane.asmamaw@ugent.be

References

[1] Alemu, B., and Kidane, D., 2014. The implication of integrated watershed Management for Rehabilitation of degraded lands: Case study of Ethiopian highlands. Journal of Agriculture and Biodiversity Research, *3, 6, 78-90.*

[2] Kidane, D., Temesgen, M., and Abdelkadir, A., 2012. Effect of winged Subsoiler and traditional tillage integrated with Fanya Juu on selected soil physico-chemical and soil water properties in the northwestern highlands of Ethiopia. East African Journal of Sciences, *6 (2), 105-116.*

[3] Descheemaeker, K., Nyssen, J., Rossi, J., Poesen, J., Haile, M., Raes, D., Muys, B., Moeyersons, J., Deckers, J. 2006. Sediment deposition and pedogenesis in exclosures in the Tigray highlands, Ethiopia. *Geoderma. 132: 291-314.* DOI: 10.1016/j.geoderma.2005.04.027

[4] Kidane, D., 2014. Conservation tillage implementation under rainfed agriculture: Implication for soil fertility, green water management, soil loss, and grain yield in the Ethiopian highlands. International Journal of Agricultural Sciences, *4 (9), 268-280.*

[5] Kidane, D., and Alemu, B., 2015. The effect of upstream land use practices on soil Erosion and sedimentation in the upper Blue Nile Basin, Ethiopia. Research Journal of Agriculture and Environmental Management, *4(2),* 055-068.

[6] Chemeda, S., 2007. Soil Erosion Hazard Assessment Using USLE Model: A CASE STUDY OF Legedadi & Dire Reservoir Catchment. www.aau.electroniclibrary. Accessed on November 13/2014.

[7] Haile, E., 2007. Land use/ Land cover Dynamics and Erosion Susceptibility Analysis: a Case Study of Basso Werena Wereda, Ethiopia. www.aau.electronic library accessed in January 2015.

[8] Hurni, H., 1985. Erosion – Productivity – Conservation Systems in Ethiopia, in Proceedings of the 4th International Conference on Soil Conservation, Maracay, Venezuela, 654-674.

[9] Hurni, H., 1988. Degradation and conservation of the resources in the Ethiopian highlands. Mountain Research and Development, *8(2/3), 123-130.* https://doi.org/10.2307/3673438

[10] FAO 1986. Ethiopian Highland Reclamation Study.Vol. 1 and 2. FAO. Rome, Italy.

[11] Kidane, D., 2015. A critical review of integrated river basin management in the upper Blue Nile river basin: The case of Ethiopia, International Journal of River Basin Management, DOI: 10.1080/15715124.2015.1013037.

[12] Lal, R., 1994. Soil Erosion Research Methods. St. Luice Press, USA, 12

[13] Nyssen, J., Govaerts, B., Araya, T., Cornelis, W M., Hans, B., Haile, M., Sayre, K., Deckers, J., 2011. The use of the *marasha* ard plows for conservation agriculture in northern Ethiopia. Agronomy and Sustainable Development, *31, 287-297.* DOI: 10.1051/agro/2010014

[14] Nyssen, J., W., Clymans, K., Descheemaeker, J., Poesen, I., Vandecasteele, M., Vanmaercke, A., Zenebe, M. Van Camp, M., Haile, N., Haregeweyn, J., Moeyersons, K., Martens, T., Gebreyohannes, Deckers, J., Walraevens, K., 2010. Impact of soil and water conservation measures on catchment hydrological response a case in north Ethiopia. *Hydrological Processes, 24(13),1880-1895.* https://doi.org/10.1002/hyp.7628

[15] Nyssen, J., Poesen, J., Moeyersons, J., Haile, M., Deckers, J., 2008. Dynamics of soil erosion rates and

controlling factors in the northern Ethiopian highlands towards a sediment budget. *Earth Surface Processes and Landforms, 33(5), 695-711.*https://doi.org/10.1002/esp.1569

[16] Ji, C.Y., 2008. Land degradation in Central Asia. Central Asian Countries Initiative for Land Management Multi-country Partnership Framework Support Project. Revised Draft Final Report. www.adb.org/documents/.../land degradation-central Asia.pdf. Accessed on November 5.

[17] Eswaran, H., Lal, R., and Reich, P.F., 2001. Land degradation: An overview In responses to land degradation. 2nd International Conference on Land Degradation and Desertification. Oxford Press, New Delhi, India.

[18] Frank, E., 2000. Gender, Agricultural Development, and Food Security in Amhara, Ethiopia: The Contested Identity of Women Farmers in Ethiopia. USAID / Ethiopia

[19] Haregeweyn, N., Tsunekawa, A., Poesen, J., Tsubo, M., Meshesha, DT., Fenta, AA., Nyssen, J., Adgo, E., 2017. Comprehensive assessment of soil erosion risk for better land use planning in river basins: A case study of the upper Blue Nile River. Science of the Total Environment 574 (2017) 95-108. https://doi.org/10.1016/j.scitotenv.2016.09.019

[20] Bogale, A., Aynalem, D., Adem, A., Mekuria, W., Tilahun, S., 2020. Spatial and temporal variability of soil loss in gully erosion in the upper Blue Nile basin, Ethiopia. Applied Water Science, 10:106. DOI: 10.1007/s13201-020-01193-4

[21] GEF 2001. Global Action on sustainable land management. Washington DC. The USA. www.gefweb.org accessed on December 17/2014.

[22] Zeleke, G., 2000. Landscape dynamics and soil Erosion process modeling in the north-western Ethiopian highlands. Ph.D. thesis. Center for Development and Environment.

[23] Hurni, H., Abate, S., Bantider, A., Debele, B., Ludi, E., Portner, B., Yitaferu, B., Zeleke, G., 2010. Land degradation and sustainable land Management in the Highlands of Ethiopia. In Hurni, H. and U. Wiesmann (eds.) (2009) Global Change and Sustainable Development: A Synthesis of Regional Experiences from Research Partnerships. University of Bern, Geographica Bernensia, Perspectives of the Swiss National Centre of Competence in Research (NCCR) North-South Vol. 5. Bern, CDE. 187-207.

[24] Muche, H., Temesgen, M., Yimer, F., 2013. Soil loss prediction using USLE and MUSLE under conservation tillage integrated with 'fanya juus' in Choke Mountain, Ethiopia. *International* Journal of Agricultural Science, *3(10), 046-052.*

[25] Outhman, S., Ahmed, A. H., Ahmed, N., Abdullah, A., 2006. Land Degradation risk Assessment in the Palestinian Central Mountains Utilizing Remote sensing and Geographic Information System (GIS) techniques. www.unu.edu/env/plec/1-degrade/index-toc accessed on November 13/2014.

[26] Tilahun, SA., Guzman, CD., Zegeye, AD., Ayana, EK., Collick, AS., Yitaferu, B., Steenhuis, TS., 2014. Spatial and Temporal Patterns of Soil Erosion in the Semi-Humid Ethiopian Highlands: A Case Study of Debre Mawi Watershed Nile River Basin. Springer, Berlin, pp. 149-163.

[27] Moges, A., Holden, N., 2008. Estimating the rate and consequences of gully development, a case study of Umbulo catchment in southern Ethiopia. Land Degradation and

Development 19(5):574-586. DOI: 10.1002/ldr.871

[28] Bewket, W., 2003. Towards Integrated watershed management in Highland Ethiopia: the Chemoga watershed case study. Tropical Resource Management Papers, No. 44. Wageningen University, Wageningen

[29] Woody Biomass in 1996. Atlas of Woody biomass inventory and strategic planning project. Amhara National Regional State Addis Ababa, Ethiopia. The Federal Democratic Republic of Ethiopia, Ministry of Agriculture.

[30] Wischmeier, W.H., Smith, D.D., 1978. Predicting Rainfall Erosion Losses: A Guide to Conservation Planning. United States Department of Agriculture, Washington, D.C

[31] Renard, K.G., Foster, G.R., Weesies, G.A., Mc Cool, D.K., and Yoder, D.C., 1979. Predicting soil erosion by water: A guide to conservation planning with the revised USLE. USDA Hand Book No. 703, USDA, Washington, D.C.

[32] Anderson, 1976. Land use and land cover classification system. Geological Survey Professional Paper 964, Washington, D. C., 1976

[33] Jensen, J. R., 1996. Introductory digital image processing. New Jersey, Trentice Hall, 1996.

[34] Desmet and Govers, 1996. A GIS Procedure for Automatically Calculating the USLE LS Factor on Topographically Complex Landscape Units.

[35] Yitaferu, B.,2007. Land degradation and options for sustainable land management in the Lake Tana Basin (LTB) Amhara region, Ethiopia. Center for Development and Environment (CDE) Geographisches Institute Universitat Bern, Bern.

[36] Assen, M., 2005. Land Use Covers Three Decades in the Jelo Micro-Catchment, Chercher Highlands, Ethiopian. Ethiopian Journal of Natural resources.

[37] Bayabil, H., Tebebu, TY., Stoof, C., Steenhuis, T., 2015. Spatial and temporal runoff processes in the degraded Ethiopian highlands: The Anjeni watershed. Hydrology and Earth System Sciences Discussions 12(4):4387-4411. doi:10.5194/hessd-12-4387-2015

[38] Tsegaw, T., 2007. Assessment of Land use/Land cover Dynamics and Soil Erosion Estimation for Sustainable Management of Land Resources. A Case study in Gozamine Wereda, Amhara Region, North Central Ethiopia www.aau.electroniclibrary. Accessed on November 10, 2014.

[39] Morgan, R.P.C., 1995. Soil erosion and Conservation. Third edition. *National Soil Resources Institute, Cranfield University.* New York Black well publishing. www.blackwellpublishing.com accessed on November 10/2014. ISBN: 9781405144674

[40] Hurni, H.,1993. Land degradation, famine and land resource scenarios in Ethiopia: World Soil Erosion and Conservation, edited by Pimentel, D., Cambridge University Press, Cambridge, UK. pp 27-61. DOI:10.7892/boris.77567

[41] Araya, T., Cornelis, WM., Nyssen, J., Govaerts, B., Gebregziabher, T., Oicha, T., Getnet, F., Raes, D., Haile, M., Saire, KD., Deckers, J.,2010. Impact of conservation agriculture on runoff, soil loss, and crop yield on a Vertisol in the northern Ethiopian highlands. 2010 19th World Congress of Soil Science, Soil Solutions for a Changing World 1-6 August 2010, Brisbane, Australia. Published on DVD. DOI:10.13140/2.1.5191.5042

Chapter 4

Determination of the Most Priority Conservation Areas Based on Population Pressure and Erosion Hazard Levels in Lesti Sub-Watershed, Malang Regency, Indonesia

Andi Setyo Pambudi

Abstract

In a watershed, the Erosion Hazard Level (EHL) is usually associated with erosion rate and existing soil solum. In Lesti Sub-Watershed, erosion rate increases every year due to erosivity factor, erodibility, the length and slope, as well as crop factor and land conservation action. Analysis of erosion associated with population pressure has not been much discussed in the Lesti Sub-Watershed. This topic needs to be explored given that the erosion rate that affects sedimentation in the Sengguruh Reservoir, as an outlet of the Lesti Sub-Watershed, cannot be separated from the population activity therein. The population activity and the choice of use of land suppress the land so that it affects the carrying capacity of the watershed. Measuring land strength is usually based on the value of existing population pressure and its effect on vulnerability or erosion hazard level. This study seeks to assess the relationship between erosion hazard level and population pressure, as well as to determine the priority conservation areas in the Lesti Sub-watershed. The research approach uses a mixed method. The results shows that from 12 sub-districts in Lesti Sub-watershed there is 1 sub-district which has high population pressure as well as severe EHL. This sub-district is the most priority area for environmental conservation.

Keywords: watershed, erosion, population, land

1. Introduction

The problem of watersheds is the problem of ecological balance related to the carrying capacity of the environment and its components [1–3]. The environment is defined as a region (region, etc.) as a boundary of economic activity, which influences the development of life in it [4, 5]. Dwelling or hydrological containers of economic activity based on environment are described as watersheds [4, 6].

The conversion of lands of an area of a watershed is due to population pressure on the land indicating there is a role for the community, both on a spatial scale and in general that affects the sustainability of natural resources [7–9]. Population pressure on this land is driven by the unbalanced rate of population growth with the availability of land resulting in increased activity and intensity on existing land or opening new land [10]. Conversion of lands without regard to topographic geological, and carrying capacity of ecosystems causes natural disasters such as landslides, floods and drought [11].

Land conversion is always associated with erosion [12]. Brantas watershed is one of the priority watersheds facing erosion problems [13, 14]. Lesti sub-watershed as part of the Brantas watershed plays a very important role in the preservation of the Sengguruh Reservoir. The Sengguruh Reservoir affects the supply of irrigation water for flood control, and generates most of the electricity in the East Java Province [15]. Erosion from upstream of the Lesti Sub-watershed sub-impacted a reduction in the storage capacity of the Sengguruh Reservoir resulting in an accelerated reduction of water storage from the original plan [16]. The interesting thing is that the upstream area of the Brantas watershed, especially the Lesti Sub-watershed is contributing a large river water flow to the downstream of the Sengguruh Reservoir [16–18].

Previous studies of erosion in the Lesti Sub-watershed show a significant upward trend in erosion rates. Yupi [19] has calculated the rate of erosion in the average of each hectare of land in the Lesti Sub-watershed, which is 30.57 tons/ha/year. The results of the Setyono and Prasetyo studies in 2012 stated that the average erosion rate in each hectare of land in the Lesti Sub-watershed was 105.763 tons/ha/year [20]. Meanwhile, the study of Ma'wa *et al.* [16] got an average erosion rate/hectare of 131,098 tons/ha/year. This research increasingly shows that areas with a high level of erosion hazard are also getting wider, especially in the current conditions.

Analysis of erosion associated with population pressure has not been much discussed in the Lesti Sub-watershed. This needs attention because the rate of erosion that affects sedimentation in the Sengguruh Reservoir as an outlet of the Lesti Sub-watershed cannot be separated from the activities of the residents therein. The activities of the population and the choice of how to use land in fact suppress the land so that it affects the carrying capacity of the watershed. Measuring land strength is usually known from the value of existing population pressures and their effects on vulnerability or the level of erosion hazard.

Research related to erosion in the Lesti Sub-watershed so far has only been influenced by physical factors of the watershed such as slope, vegetation, and soil erodibility [21]. Linkages between population pressures and the extent of the erosion hazard area in determining the most priority areas for conservation are rare [22–26]. The linkage and determination of the most priority areas for environmental conservation are interesting things to be studied further based on environmental science.

2. Materials and methods

2.1 Time and location

The time to complete the research was 12 months from conceptualization, data collection, data analysis and report writing. The research location is limited to the Lesti Sub-watershed as one of the upstream Brantas watersheds (**Figure 1**).

Administratively, the Lesti Sub-watershed is located in Malang Regency with the total area of the sub-watershed is 64,740.84 ha. The research sites cover 12 sub-districts, namely Poncokusumo, Tirtoyudo, Ampelgading, Turen, Wajak, Dampit,

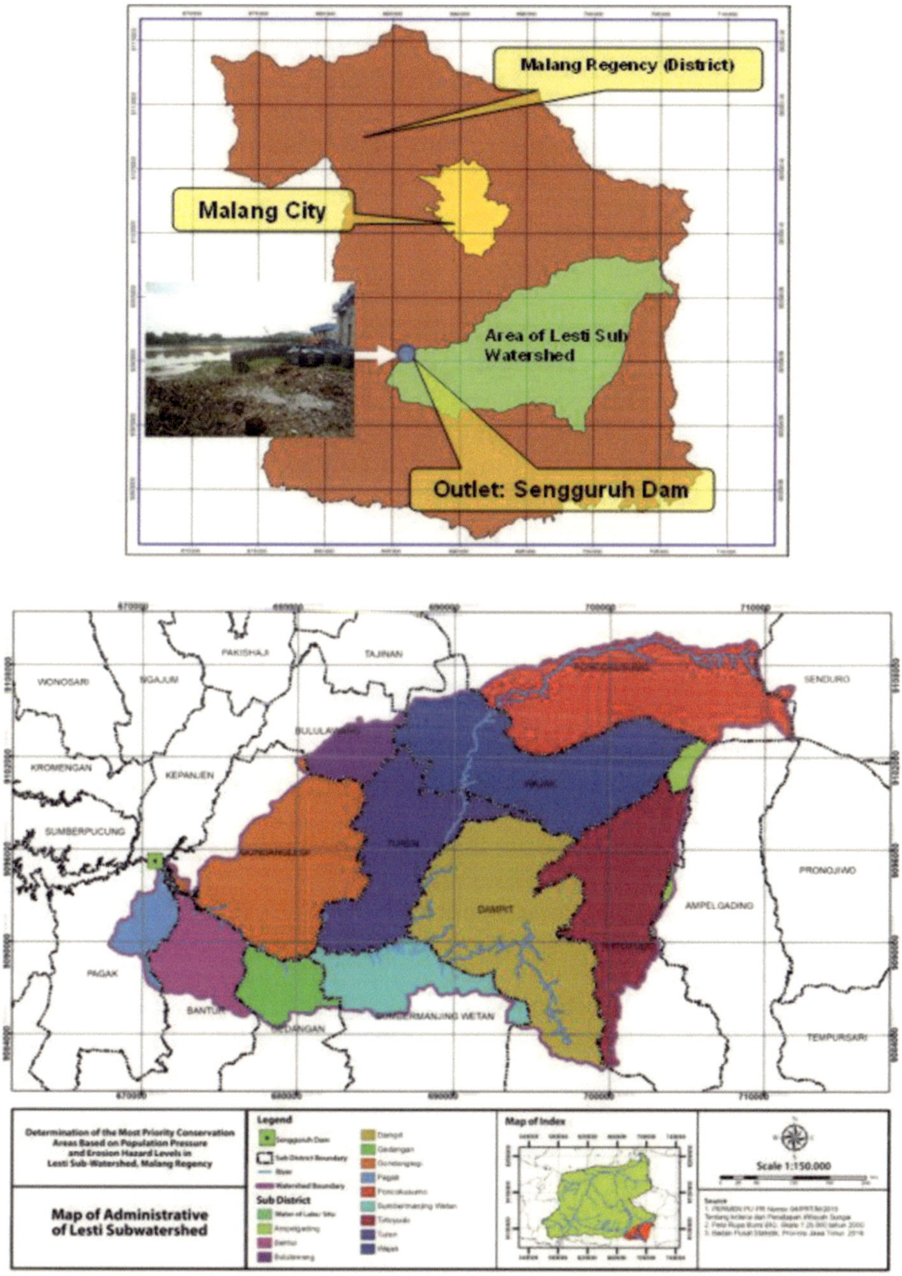

Figure 1.
Study location: Lesti sub watershed, Malang District – East Java Province, Republic of Indonesia.

Bululawang, Sumbermanjing Wetan, Pagak, Gondanglegi, Gedangan, and Bantur Sub-district. The limitation of the study area starts from the headwaters of the Lesti Sub-watershed in Poncokusumo Sub-district to the Sengguruh Reservoir outlet.

2.2 Materials and tools

This research uses several secondary data from related institutions such as: 1) rainfall data in the last 10 years; 2) the latest land use and soil data in 2018 issued by the Office for Watershed Management and Protection Forest Brantas [27] in East

Java, Indonesia; 3) contour spatial data (issued by Indonesia Spatial Information Board), soil type, slope and plant management factors and conservation measures. In addition, several tables were agreed upon by experts from previous researchers. Some secondary data related to the agricultural sector and population from agencies such as the Central Statistics Bureau (BPS) of Indonesiaand the Ministry of Agriculture of Indonesia are also needed, particularly to analyze population pressure in the Lesti Sub-watershed.

2.3 Research methods

The approach used is a mixed method with the population pressure analysis method using the Otto Soemarwoto [28] formula and the projected population growth using a geometric formula. The erosion rate calculation is analyzed using the MUSLE method with the support of Geographic Information System tools [29]. To calculate the erosivity of surface runoff as part of the MUSLE method, a modified rational formula is used. The software used is Arc GIS 10.3, and Microsoft Excel 2019. The level of erosion hazard is obtained by overlaying the erosion rate map analysis results with the soil solum map in the Lesti Sub-watershed in Office for Watershed Management and Protection Forest Brantas [27]. The choice of environmentally sound conservation priority areas based on the results of population pressure analysis >1 that intersects with the level of erosion hazard that is heavy/very heavy at the sub-district scale.

The definition of population pressure on land is a comparison of the number of people with a minimum area of land to live properly [28] (**Figure 2**).

Ideal population pressure is one that can still adjust the carrying capacity of the land. Carrying capacity of land itself is the ability of the environment to support life. The higher the percentage of land that can be used for agricultural land, the greater the carrying capacity of the land [28].

Ariani *et al.*, 2012, stated that the Population Pressure value <1 indicates that there was no population pressure or that the area was still able to meet the population's living needs in more than adequate numbers. TP value equal to 1 means that the area is still able to meet the living needs of its inhabitants appropriately. TP value is greater than 1, meaning that there has been a population pressure on the

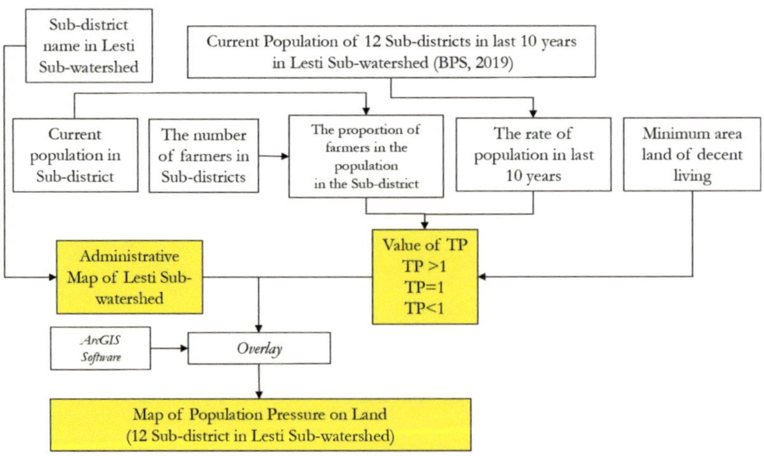

Figure 2.
Flow analysis of population pressure on land in Lesti sub-watershed, Malang District – East Java Province, Republic of Indonesia.

land in an area so that it is unable/able to meet the living needs of its residents properly.

Population pressure on land is calculated by the formula Otto Soemarwoto [28] as follows:

$$TP = Z \text{ x} \frac{\text{f } Po \ (1+\text{r})^{\text{t}}}{\text{L}} \tag{1}$$

where
TP = Population Pressure
L = Total area of agricultural land
Z = Minimum land area per farmer to be able to live properly
Po = Total population of the initial year
F = Proportion of farmers in the population (%)
T = Time span in years
R = The average population growth rate per year

The minimum land area of each farmer to be able to live properly (Z value) is calculated based on the formula as follows:

$$Z = \frac{(0.25\,LSI_2) + (0.5LSI_1) + (0.5LST) + (0.76LLK)}{(LSI_2 + LSI_1 + LST + LLK)} \tag{2}$$

where
Z = Minimum land area per farmer to be able to live properly (ha)
LST = The area of rain-fed rice fields (ha)
LLK = Dry land area (ha)
LSI_1 = The area of irrigated rice field once a year harvest (ha)
LSI_2 = The area of Irrigated rice field from twice a year harvest (ha)

The proportion value of farmers in the population (f) is obtained from the formula submitted by Soemarwoto [28], namely:

$$\text{f} = (\text{Number of farmers / Total population}) \text{ x } 100\% \tag{3}$$

The population growth rate is calculated using the geometric formula as follows:

$$\text{Pt} = \text{Po}(1+\text{r})^{\text{t}} \tag{4}$$

where
r = Population growth rate
t = The time period, which is stated in years
Pt = Total population in the year t
Po = Total population of the initial year

In order for calculating the rate of erosion using the formula of the MUSLE (Modify Universal Soil Loss Equation) in **Figure 3** and below

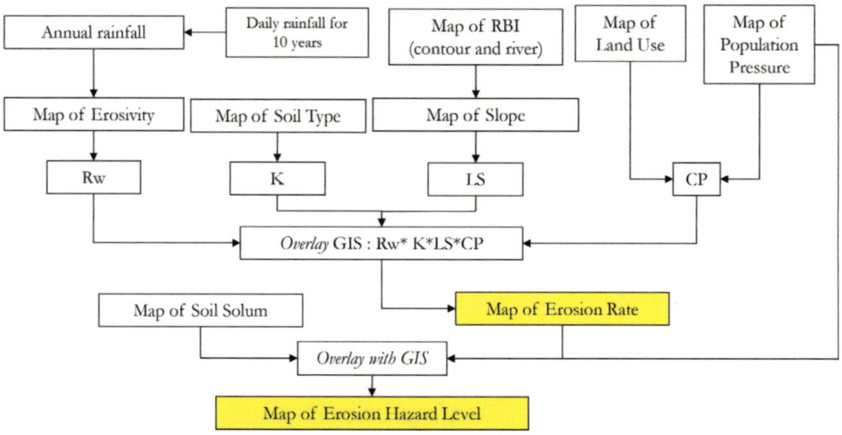

Figure 3.
Flow analysis of erosion rates and erosion Hazard levels in Lesti sub-watershed.

$$\mathbf{A} = \mathbf{R_W} \times \mathbf{K} \times \mathbf{LS} \times \mathbf{CP} \tag{5}$$

$$\text{where } R_W = 9{,}05\left(V_O \times Q_p\right)^{0{,}56}$$

Note:

- A = Erotion Rate (ton/ha/tahun)
- R_W = Surface runoff erosivity index *(run-off)*
- K = Soil erodibility factor
- LS = Slope factor
- CP = Factors of land use and land management
- V_O = Surface runoff volume (m^3)

Runoff discharge (Qp) is calculated in relation to the surface runoff erosivity (Rw) as part of the MUSLE method erosion estimation formula. In order to get runoff discharge data, several steps are needed, namely: 1) Determining the Flow Coefficient (C); 2) Determine the Concentration Time (Tc), Reservoir Coefficient (Cs) and Rainfall Intensity (I); 3) Calculate runoff discharge and describe it in the form of Run-off Discharge Distribution Maps (surface runoff) in various times with ArcGIS 10.3.

The next step is the calculation of runoff discharge. Determination of the amount of runoff discharge is done through overlays with ArcGIS software. This analysis is done through geoprocessing analysis on ArcGIS 10.3 software. The data used are in Lesti Sub-watersheds (Coefficients Cs and I), land use maps (for C Coefficient). The formula used is the modified rational runoff equation, namely:

$$Q = 0{,}00278.\ Cs.\ C.\ I.\ A \tag{6}$$

Sub-districts in Lesti sub-watershed	Total population	Number of farmers	Proportion of farmers	Growth rate population	Minimum land area decent life	Agricultural land area (Ha)	Value of population pressure	Criteria
Poncokusumo	26.221	24.460	0,93	1,53	0,17	4.226,381	0,97,542	< 1
Wajak	74.121	66.292	0,89	1,20	0,19	4.621,481	0,87,528	< 1
Dampit	108.914	89.087	0,82	1,50	0,19	8.361,963	**361,627**	**> 1**
Tirtoyudo	44.121	28.991	0,66	1,44	0,17	3.029,741	0,86,021	< 1
Sumbermanjing Wetan	24.739	15.099	0,61	1,47	0,19	1.548,180	0,55,234	< 1
Turen	107.607	61.445	0,57	1,68	0,16	3.713,927	**368,583**	**> 1**
Bululawang	12.282	4.927	0,40	0,72	0,16	209,196	0,00427	< 1
Gondanglegi	82.052	57.984	0,70	1,50	0,16	5.444,617	**195,847**	**> 1**
Ampelgading	14.823	9.084	0,69	1,24	0,16	307,824	0,13,464	< 1
Gedangan	12.032	5.043	0,42	0,19	0,26	1.329,656	0,00001	< 1
Bantur	20.384	13.051	0,64	0,66	0,26	1.757,160	0,01192	< 1
Pagak	7.683	7.123	0,93	1,49	0 26	1.082,391	0,38,289	< 1

Source: Analysis Result, 2019.

Table 1.
Level of population pressure on land (bold values showed Area with high populationpressure in Lesti-Sub-Watershed).

3. Results and discussion

Based on the existing formula related to calculation analysis for population pressure, the results are as shown in **Tables 1** and 2 and **Figures 4–6**. These results are a combination of spatial calculations with Arc GIS and calculations using Microsof Excel.

The conditions shown in **Table 1**. indicate that there are 3 sub-districts that have level of population pressure on high land, namely Dampit Sub-district, Turen

Number.	Sub-sub watershed	Area (ha)	Result of erosion rate (A) (ton/ha/year)
1	1	2244,760	60,897,267
2	2	1272,640	223,693,161
3	3	2585,000	66,900,000
4	4	4662,400	1,487,060,443
5	5	171,960	7416,818
6	6	3090,400	659,022,308
7	7	317,520	34,850,582
8	8	2945,280	833,252,748
9	9	140,480	3113,887
10	10	2574,120	11,963,945
11	11	4081,720	17,377,677
12	12	2224,800	1,056,466,594
13	13	1464,680	586,835,010
14	14	1653,560	85,246,379
15	15	2388,720	108,713,978
16	16	280,080	6605,734
17	17	1828,480	71,480,000
18	18	4787,960	35,165,209
19	19	1360	521,939
20	20	4800	933,119
21	21	2781,720	648,610,551
22	22	192,040	19,518,057
23	23	1613,120	429,947,887
24	24	1898,440	254,515,524
25	25	1412,760	81,638,228
26	26	2285,200	182,243,500
27	27	2224,520	14,097,787
28	28	1674,480	998,110,000
29	29	4468,480	283,810,000
30	30	2922,560	776,230,000
31	31	4546,800	915,280,000
Total		64,740,84	9,961,518,329

Table 2.
Calculation result related to erosion rate (generated by GIS) of Lesti subwatershed, Malang District – East Java Province, Republic of Indonesia.

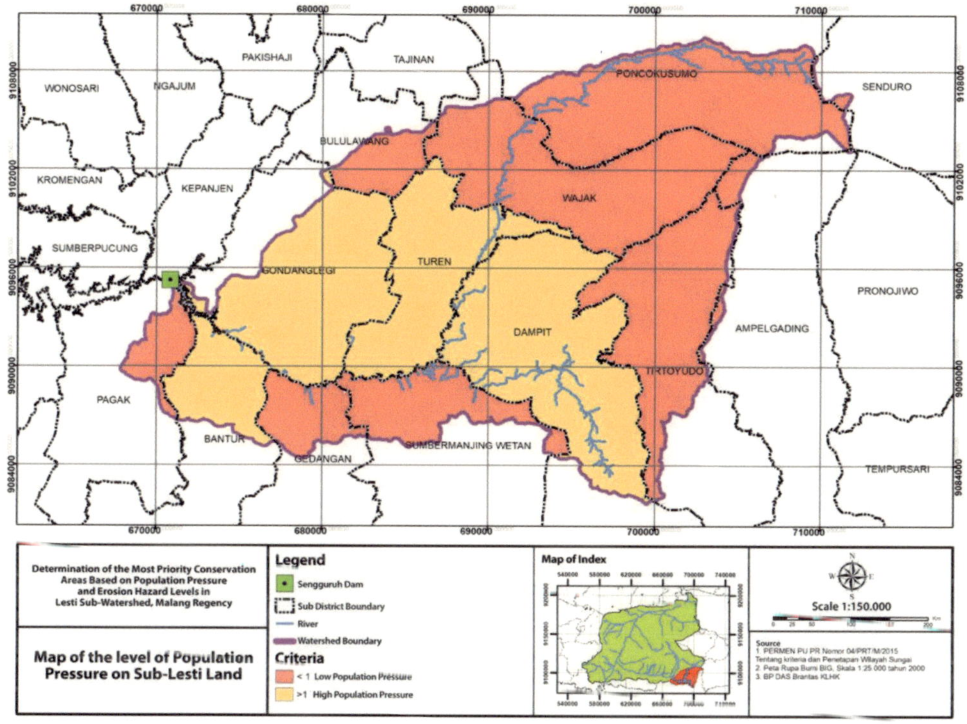

Figure 4.
Map of population pressure conditions in 12 sub-districts of Lesti sub-watershed, Malang District – East Java Province, Republic of Indonesia.

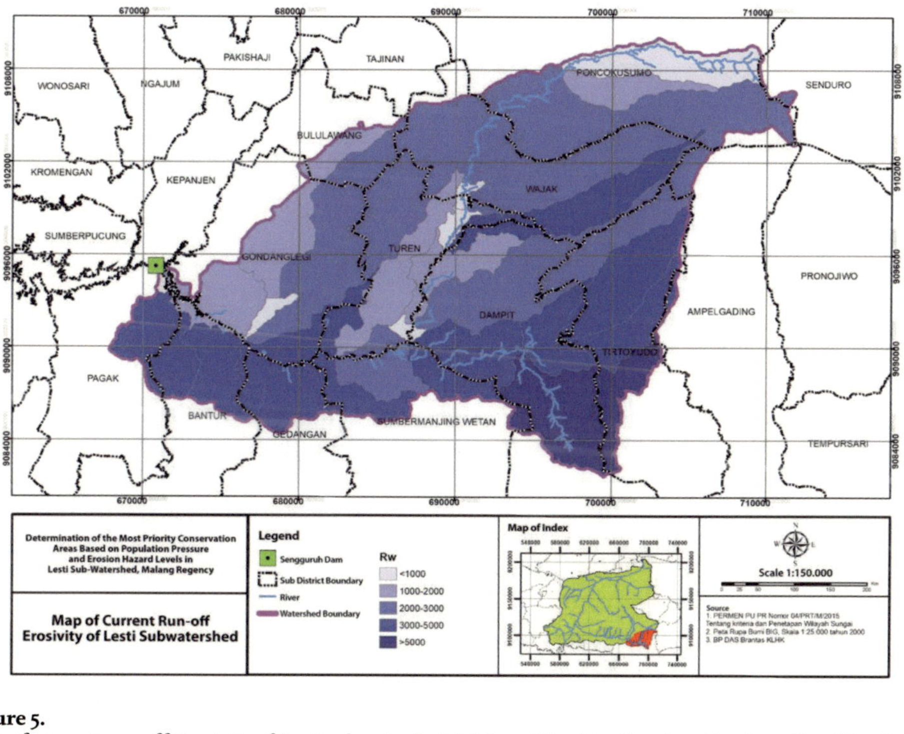

Figure 5.
Map of current run-off Erosivity of Lesti subwatershed, Malang District – East Java Province, Republic of Indonesia.

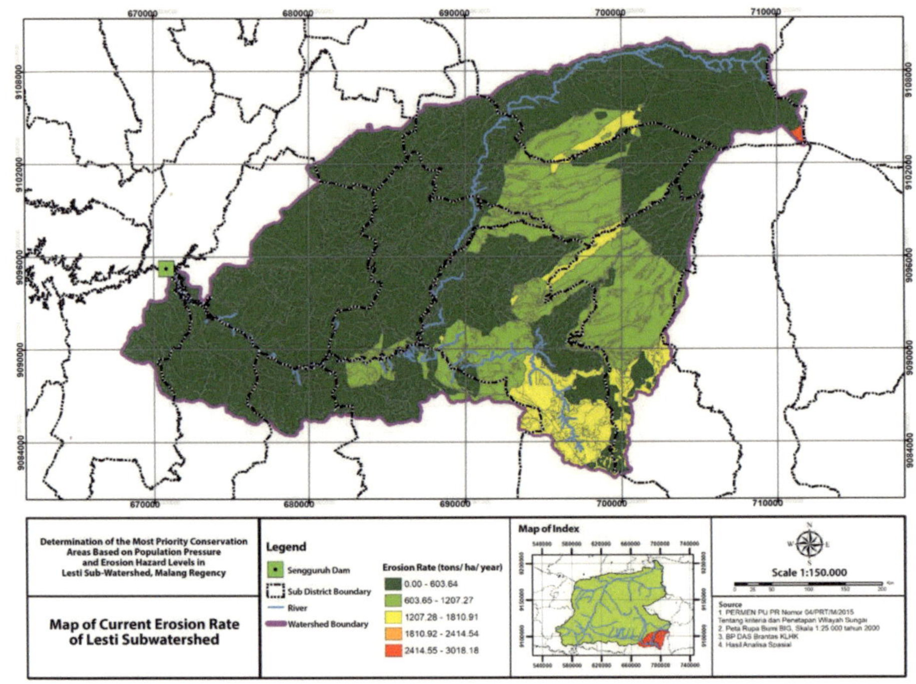

Figure 6.
Map of current erosion rate of Lesti subwatershed, Malang District – East Java Province, Republic of Indonesia.

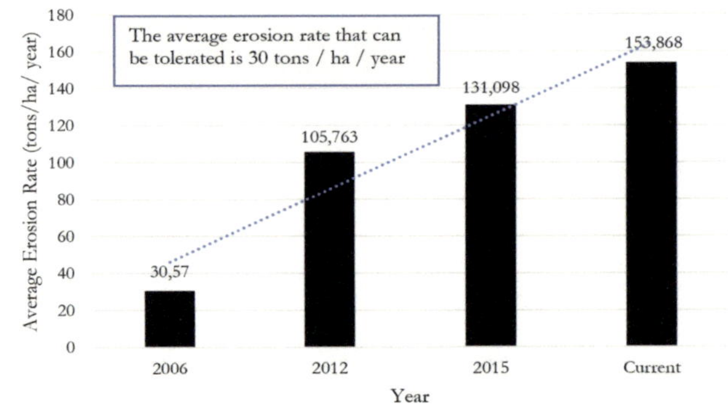

Figure 7.
Average erosion rate of Lesti sub-watershed from 2006 – Present.

Sub-district and Gondanglegi Sub-district. If looked at the current land use, several sub-districts identified as having a Population Pressure (TP) > 1 are on open land.

Based on the results of the above calculations, it is known that the total recent erosion rate in the Lesti sub-watershed is 9,961,518,329 tons/ha/year. Considering that the value of the sediment delivery ratio in the Lesti sub-watershed is 8.247%, the amount of sediment in the sub-watershed is 821,556.3 tons/ha/year.

Meanwhile, with the Lesti sub-watershed area of 64,740.84 ha, it can be calculated that the current average erosion rate in each ha of land in the Lesti sub-watershed is 153,868 tonnes/ha/year (exceeding the tolerable erosion rate of 30 tonnes/ha/year). Previous research results from Yupi [19] stated that the average erosion rate in the Lesti sub-watershed was 30.57 ton/ha/year, and Setyono and Prasetyo's research (2012) was 105.763 tonnes/ha/year. Meanwhile Ma'wa et al. In 2015, the average erosion rate was 131.098 ton/ha/year. **Figure 7**. Based on the results of calculations carried out by the author which states that the current erosion rate is 153.868 tonnes/ha/year, it can be said that there has always been an increase in the erosion rate of the Lesti Sub-watershed since the last 14 years so that better conservation management is needed.

The erosion rate calculation in the Lesti Sub-watershed is used as a basis for obtaining the extent and information on the Erosion Hazard Level Category (TBE) **Tables 3** and **4**. The values estimate the maximum soil loss that will occur on a land. Spatially, the TBE map makes it easy to see the condition of certain areas as conservation priority areas. TBE map is obtained by overlaying the current erosion rate map, behavior map and population pressure map with the soil solum map in the Lesti Sub-watershed **Figure 8**.

Some of the sub-districts identified as having the highest area of Erosion Hazard Levels marked in red on the map are in Wajak Sub-district, Tirtoyudo Sub-district, Dampit Sub-district, Sumbermanjing Wetan Sub-district, Gedangan Sub-district and Bantur Sub-district. As is known Dampit Sub-district, Turen Sub-district and Gondanglegi Sub-district have TP values >1, which means that there has been a population pressure on the land in an area so that it has not been able to meet the needs of its population properly **Figure 9**.

From the standpoint of environmental science, it can be said that erosion which is usually seen from the aspect of carrying capacity of the environment, also has a strong connection with social and economic aspects in the form of pressure. Based on TBE and TP analysis, it was found that 1 sub-district had slices, namely Dampit Sub-district. It is recommended that environmental conservation directives focus

No	ID Solum	Depth	Class of Solum soil	Area (m^2)	Area (Ha)	Percentage (%)
1	A	> 90 cm	Deep	561,419,204,2	56,141,92,042	86,72
2	B	60–90 cm	Medium	68,067,150,77	6806,715,077	10,51
3	C	30–60 cm	Shallow	9,930,132,548	993,0132548	1,53
4	D	< 30 cm	Very shallow	7,991,912,474	799,1,912,474	1,23
Total				647,408,400,00	64,740,84	100,00

Source: Analysis Result, 2019.

Table 3.
Data of solum soil depth in Lesti sub-watershed.

No	Erosion hazard level	Area (m^2)	Area (Ha)	Percentage (%)
1	Very light	115,142,109	11,514,21	17,79
2	Light	113,345,070	11,334,51	17,51
3	Medium	97,183,967	9718,40	15,01
4	Heavy	69,542,700	6954,27	10,74
5	Very heavy	252,194,553	25,219,46	38,95
Total		**647,408,400**	**64,740,84**	**100,00**

Source: Analysis Result, 2019.

Table 4.
Percentage of erosion hazard level of current Lesti sub-watershed.

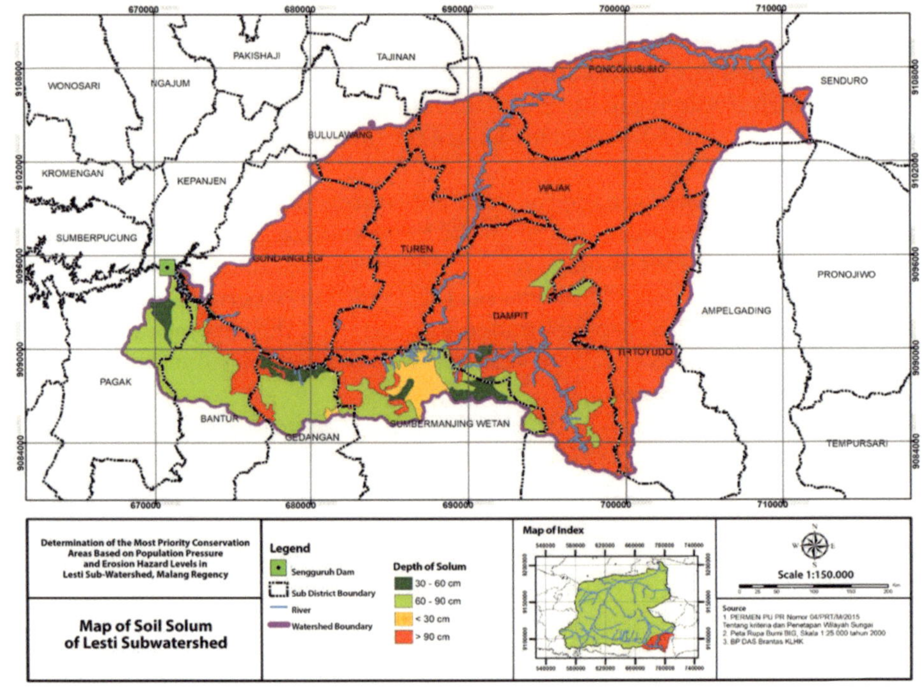

Figure 8.
Map of soil solum of Lesti subwatershed.

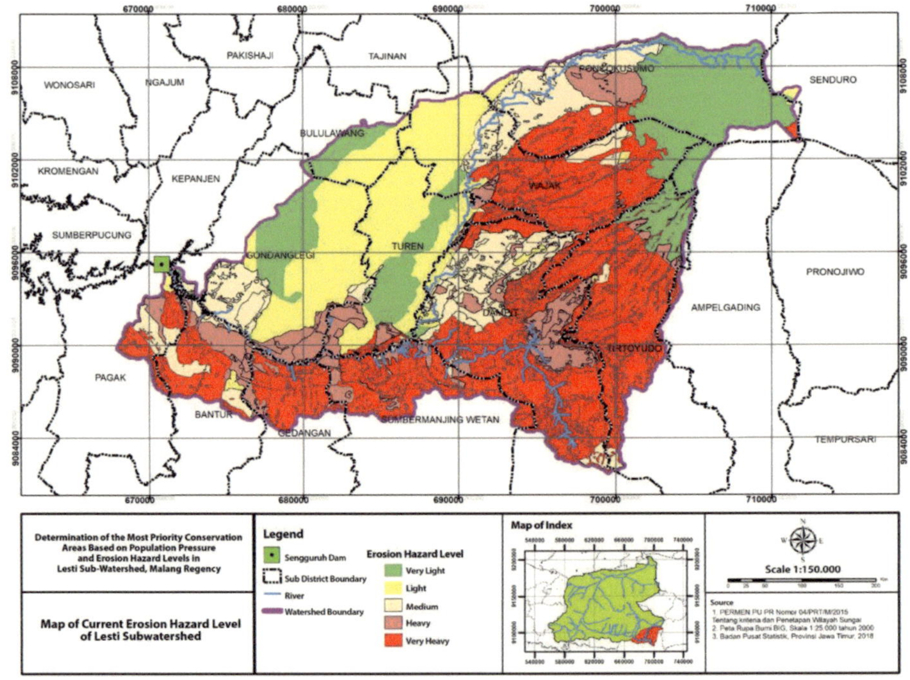

Figure 9.
Map of current erosion hazard levels of Lesti sub-watershed.

on the sub-district through the application of technical, vegetative, agronomic, land and water civil conservation as well as a combination involving the community and in accordance with local conditions.

4. Conclusion

Based on the results of data analysis and discussion, it can be concluded that the Lesti Sub-watershed, there is a correlation between population pressure and the current choice of land use, which results in erosion vulnerability. At high population pressure (> 1) in general is directly proportional to the erosion-prone land use conditions such as settlements, dry land fields and open land. Based on environmental science, the government needs to balance economic, social and environmental needs in several regions. Priority for conservation is prioritized in Dampit Sub-district because it is an area with high TP slices and heavy TBE.

The recommendation that can be given to this sub-district is the provision of subsidies or incentives by the government for people who want to carry out agricultural efforts with conservation principles. This is to reduce the income gap because in some cases of agricultural output will decrease when applying the principle of environmental conservation. In the social aspect, efforts are needed to involve the community with their local wisdom to carry out conservation efforts, both technical civil, agronomic and vegetative so that there is a sense of ownership of government programs undertaken to prevent erosion in the upstream watershed.

Acknowledgements

The authors gratefully thanks to Director Agustin Arry Yanna, SS, MA; Prof. Drs. Raldi Hendrotoro Seputro Koestoer, M.Sc., Ph.D.; Prof. Dr. Ing. Ir. Dwita Sutjiningsih, Dipl. HE; Dr. Muchamad Saparis Soedarjanto, M.T., S.Si; Rahmad Junaidi, ST, MT; Sri Hidayati. S.Si, M.Han; Dr. Hayati Sari Hasibuan, ST, MT and all parties who assist in writing, both technical and non-technical.

Author details

Andi Setyo Pambudi
Postgraduate School of Environmental Sciences, University of Indonesia, Salemba Campus, Jl. Salemba Raya No. 4, Kampus UI Salemba, Kenari, Kec. Senen, Kota Jakarta Pusat, Indonesia

*Address all correspondence to: andisetyopambudi@gmail.com; andi.pambudi@bappenas.go.id

References

[1] Asdak C. *Hidrologi dan Pengelolaan Daerah Aliran Sungai. Edisi* Kelima (Revisi). Yogyakarta: Gadjah Mada. University Press; 2010

[2] Chaidar AN, Soekarno I, Wiyono A, Nugroho J. Spatial analysis of erosion and land criticality of the upstream Citarum watershed. *International Journal of GEOMATE,* September, 2017, Volume13. In: doi:https://doi.org/10.21660/2017.37.34572. 2017

[3] Williams JJ. *Sediment – Yield prediction with universal equation using runoff energy factor*. Environmental Science. 1975:12-14 https://www.semanticscholar.org/paper/Sediment-yield-prediction-with-Universal-Soil-Loss-Equation-Williams

[4] Common M, Stagl S. *Ecological Economic: An Introduction*. New York: Cambridge University Press; 2005

[5] Helen B, Leggett M, Mandar T, Pareira J, Gangga A. *How Can Indonesia Achieve Water, Energy and Food Security?* Jakarta: WCS and Global Canopy Programme; 2016

[6] Miller GT, Spoolman SE. *Living in the Environment: Concepts*. Brooks/Cole, Belmont, CA (USA): *Connections and Solutions. Seventeenth edition*; 2015

[7] Cumming, G. S. (2016). The relevance and resilience of protected areas in the Anthropocene. *Anthropocene, 13, 46-56.* doi:10.1016/j.ancene.2016.03.003

[8] Mtibaa S, Hotta N, Irie M. Science of the Total Environment Analysis of the Ef Fi Cacy and Cost-Effectiveness of Best Management Practices for Controlling Sediment. Yield: A Case Study of the Joumine Watershed, Tunisia. *Journal Science of the Total Environment*. Elsevier B.V; 2018. DOI: https://doi.org/10.1016/j.scitotenv.2017.10.290

[9] Watson JEM, Dudley N, Segan DB, Hockings M. The performance and potential of protected areas. Nature. 2014;*515*(*7525*):*67-73*. DOI: 10.1038/nature13947

[10] Soemarwoto O. *Analisis Mengenai Dampak Lingkungan*. Yogyakarta: Gadjah Mada. University Press; 1999

[11] Sinukaban N. *Peranan Konservasi Tanah dan Air dalam Pengelolaan Daerah Aliran Sungai*. Jakarta: Prosiding Bunga Rampai Konservasi Tanah dan Air; 2007

[12] Arsyad S. *Konservasi Tanah dan Air*. Bogor: IPB Press; 2006

[13] Bappenas. *Kajian Pengaruh Kebijakan Konservasi Sumber Daya Air di dalam DAS Terhadap Sektor Kehutanan dan Sektor lainnya*. Jakarta: Bappenas; 2015

[14] GoI (2014). *Perpres No. 2 Tahun 2015 tentang Rencana Pembangunan Jangka Menengah Nasional Tahun 2015-2019.* Jakarta: Bappenas

[15] Pambudi AS, Moersidik SS. Conservation direction based on estimation of erosion in Lesti sub-watershed, Malang District. In: *International Proceeding IOP Conference Series: Earth and Environmental Science, Volume 399, the 1st International Seminar on Natural Resources and Environmental Management 2019*. Published under licence by: IOP Publishing Ltd; 2019. DOI: 10.1088/1755-1315/399/1/012097

[16] Ma'wa J, Andawayanti U, Juwono PT. *Studi Pendugaan sisa usia guna Waduk Sengguruh dengan pendekatan erosi dan sedimentasi*. Skripsi: Jurusan Pengairan Fakultas Teknik Universitas Brawijaya; 2015

[17] Djajasinga V, Masrevaniah A, Juwono PT. Kajian Ekonomi Penanganan

Sedimen Pada Waduk Seri Di Sungai Brantas (Sengguruh, Sutami Dan Wlingi). *Jurnal Teknik Pengairan, Volume 3, Nomor 2*. Desember 2012, hlm. 2012:*143-152*

[18] Yetti E, Soedharma D, Haryadi S. Evaluasi kualitas air sungai-sungai di kawasan DAS Brantas Hulu Malang dalam Kaitannya dengan tata guna lahan dan aktivitas masyarakat di sekitarnya. Jurnal Pengelolaan Sumberdaya Alam dan Lingkungan (JPSL). 2011;**1**(*1*): *10-15*. DOI: https://doi.org/10.29244/jpsl.1.1.10

[19] Yupi HM. *Studi model WEPP (Water Erosion Prediction Project) dalam upaya pengaturan fungsi kawasan pada Sub DAS Lesti berbasis Sistem Informasi Geografi (SIG)*. Tesis: Universitas Brawijaya; 2006

[20] Setyono and Prasetyo. (2012). Analisis Tingkat Bahaya Erosi Pada Sub DAS Lesti Kabupaten Malang Menggunakan Sistem Informasi Geografis. *Media Teknik Sipil, Volume 10, Nomor 2, Agustus 2012: 114-127.* DOI: https://doi.org/10.22219/jmts.v10i2.1786

[21] Kagoya S, Paudel KP, Daniel NL. Awareness and Adoption of Soil and Water Conservation Technologies in a Developing Country: A Case of Nabajuzi Watershed in Central Uganda. *Journal of Environemental Management*. Publisher: Springer Science+Business Media, LLC; 2017. DOI: 10.1007/s00267-017-0967-4

[22] Hardjowigeno S. *Ilmu Tanah*. Jakarta: Akademika Pressindo; 1995

[23] Koentjaraningrat. *Pengantar Ilmu Antropologi*. Rineka Cipta: Jakarta; 2009

[24] Notoadmodjo S. *Promosi Kesehatan dan Ilmu Perilaku*. Jakarta: Rineka Cipta; 2007

[25] Tambunan R. *Perilaku Konservasi pada Masyarakat Tradisional*. In: Jurnal Harmoni Sosial, Januari 2008. Volume: II; 2008 No. 2

[26] Utomo WH. *Erosi dan Konservasi Tanah*. Malang: IKIP; 1994

[27] BPDAS-HL Brantas. *Map of Lesti subwatershed, Brantas upstream watershed. Jawa Timur.* Jakarta: Kementerian Lingkungan Hidup dan Kehutanan (Indonesia's Ministry of Environment and Forestry). 2019

[28] Soemarwoto O. *Ekologi, Lingkungan Hidup dan Pembangunan*. Jakarta: Jambatan; 1985

[29] Tajbakhsh SM, Memarian H, Kheyrkhah A. A GIS-based integrative approach for land use optimization in a semi-arid watershed. Global Journal of Environmental Science Management. 2018;**4**(1):31-46. DOI: http://doi.org/10.22034/gjesm.2018.04.01.004

Chapter 5

The Impacts of Soil Degradation Effects on Phytodiversity and Vegetation Structure on Atacora Mountain Chain in Benin (West Africa)

Farris Okou, Achille Assogbadjo and Brice Augustin Sinsin

Abstract

Atacora mountain is a particular ecosystem of West Africa where soil degradation occurs. The present study assessed the impacts of physical soil degradation on vegetation in the Beninese portion of this mountain chain. Phytosociological surveys were carried out along line transects from plain to summit within 22 plots of 30 m x 30 m. Based on indicators of physical soil degradation each plot was classified into one soil degradation class (Light, Moderate, High or Extreme). Impacts on plant diversity were assessed by comparing the floristic composition of soil degradation classes with the index of similarity of Jaccard. Variations between soil degradation classes of species richness, species chorological types, species life forms and species dispersal were also tested using a discriminant analysis combined with ANOVA. The Multi-Response Permutation Procedures analysis was used to pairwise compare the soil degradation classes based on the cover data of the species lists. All soil degradation classes were dissimilar, depending on the floristic composition. Discriminant analysis and ANOVA performed on biodiversity indicators had shown that species richness, and the number of regional species, phanerophytes and sarcochory decreased along the increasing degradation gradient in contrast to the number of species with wide distribution, therophytes and sclerochory. With regard to vegetation structure, the results had shown that only moderately and highly degraded soils presented the similar vegetation type. Physical soil degradation induced modification of floristic composition, phytodiversity loss and modification of vegetation structure. These results showed that the soil degradation gradient corresponds to a vegetation disturbance gradient.

Keywords: soil degradation, phytodiversity loss, mountain chain, West Africa

1. Introduction

Land degradation has become a global problem affecting at least a quarter of all terrestrial biomes and agro-ecologies, and occurring in many low-income as well as industrialized countries [1]. Understanding and assessing the underlying processes of land degradation is important to develop suitable land management measures and policies. Land degradation involves many interrelated processes such as soil erosion, depletion of soil nutrients, loss of biodiversity, deforestation, loss of ecosystem services etc. [2].

Many studies examined the impacts of land degradation on vegetation. In general, the methodologies used consisted in statistically testing differences in certain measures of vegetation structure, biodiversity and/or ecosystem services collected over different states or intensities of degradation of a given environmental component. Some authors examined the diversity and changing composition of plant communities of different land use and land cover types under different grazing pressure intensities [3–5]. Others have addressed the difference in species diversity between forest successional stages [6, 7] or between concretion soil, sand-clay soil and *Bowal*, (considered as the final of land degradation) [8]. *Bowa*l (plural *bowé*) comes from the *fulfulde* language spoken in Guinea and refers to degraded lands found on hardened ferruginuous soils also known as ferricretes [8]. However, we are not aware of any studies that have attempted to assess the impacts on vegetation (structure and diversity) of soil degradation defined as physical soil degradation classes.

Soil is a key resource that manages the cycle of water, cycle of carbon, plant growth and distribution, fauna and geochemicals [9–11]. Soils play an important role in mountainous areas often characterized by steep slopes and shallow soils. In Benin, the mountainous Atacora region is confronted with different soil degradation processes. Increased human activities (unsustainable agriculture, livestock grazing, fuelwood and tree cutting), combined with steep slopes, shallow soils and heavy rainfall had led to soil degradation [12–16].

Into the mountainous Atacora region, previous study in Ref. [17] had examined various indicators of land degradation and found that soils could be classified into 4 soil degradation categories i.e. light, moderate, high, and extreme degradation. However, nothing is known about the impacts of soil degradation classes on vegetation. Up to now investigations about phytodiversity into the mountainous region have mainly focused on characterization of plant communities and assessment of species diversity through phytosociological surveys [18–20]. There is need to fill a gap in scientific researches and to contribute to sustainable land management in the study area by enhancing the knowledge of land degradation processes.

For the assessment of plant diversity, different methods and indices are available, including vegetation structure, floristic composition and specific richness, chorological types, life forms and dispersal types of diapores which are good indicators of the state of vegetation health [7, 21–23]. The aim of the present study was to explore the impacts of soil degradation classes on vegetation namely vegetation structure, floristic composition, species richness, chorological types, life forms and dispersal types of diapores.

2. Material and methods

2.1 Sampling data and classification of plots into soil degradation classes

Data were collected in two steps. The first step consists in the identification of sampling sites (**Figure 1**). Based on vegetation, soil and administrative map, sampling sites were chosen according to the vegetation types, the proximity to hillsides and the accessibility during rainy season. Altogether four (4) sampling sites were identified at the rate of two sites per district (Natitingou and Toucountouna). The second step consists on the data collection. Local knowledge on soil erosion was

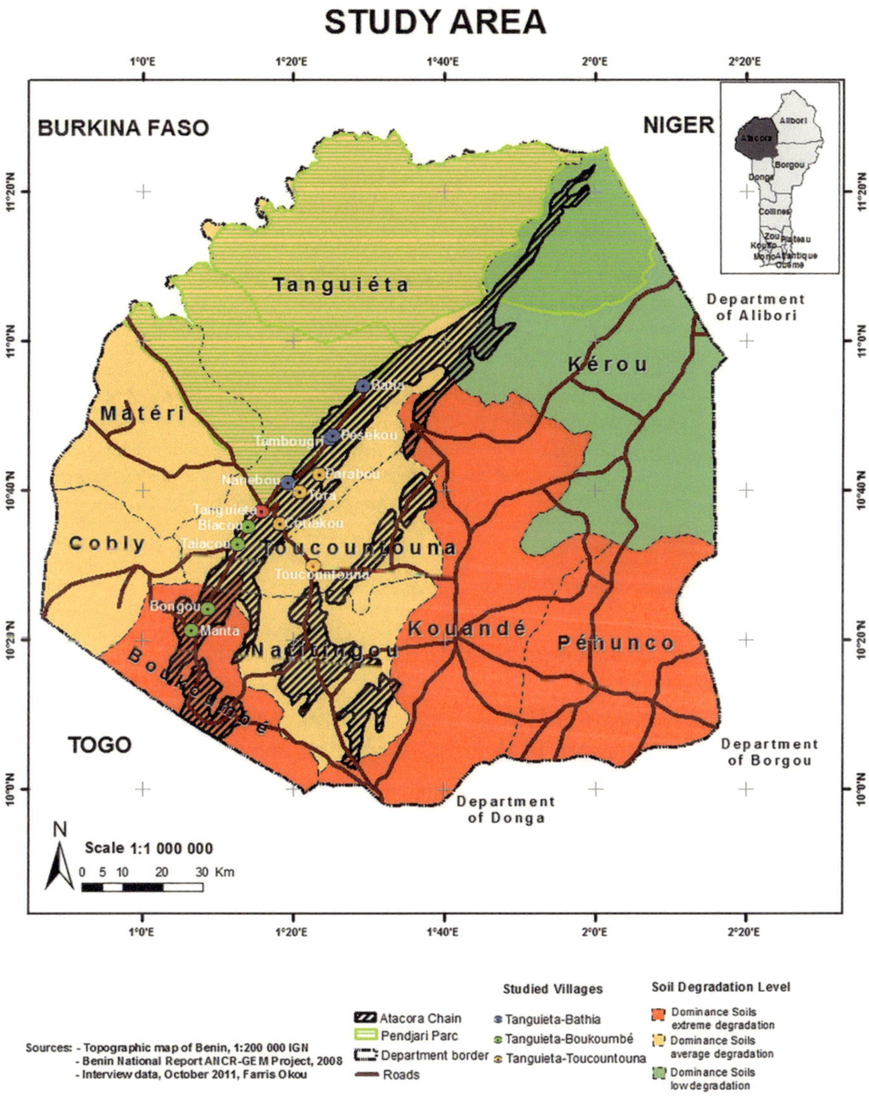

Figure 1.
Map of study area.

used in order to identify where to install the line transects. With the help of villages leaders and the guide, areas within natural vegetation, on/near mountains or hillsides where physical soil degradation occurs were identified. Within each site, one or two line transects (from plain to top) were established. At each topographical position nested sample plots (30 m x 30 m for woody layer and 10 m x 10 m for herbaceous layer) within representative and homogenous vegetation areas were installed. 22 plots of 30 m x 30 m were considered and five sub-plots each of 10 m × 10 m (four in the corner of the plot and one in its center) per plot were used.

On the basis of physical soil degradation indicators (extent of organic layer, color of topsoil, compactness of soil, presence and extent of rills, and occurrence of sheet erosion) each plot was classified visually into specific soil degradation classes. Physical soil degradation in the study area falls into four grades, namely light, moderate, high and extreme soil degradation classes described in [17]. The characteristics of each class are summarized in **Table 1**.

Soil degradation classes	Definition
Light	Soils characterized by a low level of soil compaction, few rills and no visible sheet erosion. On the topsoils black organic layer covered the entire surface and no reddish soils were observed.
Moderate	Soils characterized by compact red soils, with a thin clay crust on the surface. Sheet erosion occurred on these soils, and rills were observed on the surface. Organic layers remained as thin patches.
High	Soils red and very compact. They looked like ferricrete but remained friable. Sheet erosion occurred. Rills covered a larger surface than on the other soils and were also deep. Organic layer remained as thin patches (less than 5 cm thick) anic layer were less extended (only 23% of rod contacts).
Extreme	Soils characterized by the presence of ferricrete (rich in iron, and hard) and red soils. The presence of the ferricrete layer reduced the depth to which roots could grow. The organic layer remained only on small patches. The thickness of the organic layer rarely exceeded 10 cm. There was no visible evidence of sheet erosion, and the presence of rills was very low because of the high level of compaction.

Table 1.
Characteristics of soil degradation classes on Atacora mountain range.

2.2 Assessing impacts of soil degradation on phytodiversity

Phytosociological surveys [24] were carried out in each sample as a mean to assess the floristic composition, discriminant species, species richness, species chorological types, species life forms and species dispersal types. Woody species were collected in the plots, while herbs were carried out on the sub-plots. All species were constituted as herbaria and were subsequently determined by the National Herbarium of the University of Abomey-Calavi.

The similarities in species composition between classes of soil degradation were assessed using the index of similarity of Jaccard (1901), which is given by the formula:

$$P_j = 100 * \frac{c}{a + b - c}; \quad (1)$$

where P_j is Jaccard community coefficient, a is the number of species present in the community A, b is the number of species in the community B, and c is the number of species shared by A and B. In the study, soil degradation classes represented communities. The computation was automatically performed with the software CAP [25] on a presence/absence matrix consisting of a number of defined soil degradation classes and 133 plant species. This index has proved to be a consistently good measure of similarity for presence/absence data [26]. The values of P_j range from 0% for an absence of similarity to 100% for a complete similarity. Plant communities are dissimilar if $P_j \leq 50\%$.

Discriminants species of each degradation class was assessed and identified based on methodology as in Ref. [27]. Discriminant species of a particular group were species devoted to that group, exclusive to that group and never occurring in others groups. Dufrêne & Legendre's method produced indicator values for species within each group. These indicator values were tested for statistical significance using a randomization (Monte Carlo) technique [28]. P value of 5% was used to retain as discriminant species. All multivariate analyses were computed with PC-ORD for Windows Version 5 [28].

The impacts of soil degradation on phytodiversity were also assessed by using species richness (S), and three indexes of diversity that were developed as part of this study: the chorological index (I_C), the life forms index (I_L) and the dispersal types (of diaspore) index (I_D). The objective was to understand how biodiversity indicators vary according to soil degradation classes, i.e. along degradation gradient.

These indexes were computed on the base of two main principles. The first one was the principle of biodiversity's indicators of disturbance. Along a gradient of disturbance, there were three major types of qualitative indicators of biodiversity (chorological types, life forms and dispersal types) which evolutions (in terms of number or cover) were negatively correlated. For example, widely distributed species, therophytes and sclerochory were assumed to be more abundant/dominant in the pioneer (more disturbed) stages and this trend decreased as less disturbed stages were reached. In the contrary, the number/cover of regional species, phanerophytes and sarcochory were assumed to increase from disturbed to stable communities [21, 29, 30]. The second principle is about the ratio or relative frequency used in Ref. [31] to calculate the phytogeographical index (I_p) which made it possible to compare and classify the different plant communities according to their level of affinity with the Sudanian or Guinea-Congolian region. On this basis, the indexes were computes as:

$$I_c = \frac{S + SZ + SG}{Pt + PAL + AA + TA + PRA}; \quad (2)$$

Where I_C is the chorological index and *S, SZ, SG, Pt, PAL, AA, TA, PRA* are respectively the frequency of Sudanian, Sudano-Zambezian, Sudano-Guinean, Pantropical, Paleotropical, Afro-American, Tropical Africa and Pluri Regional in Africa species.

$$I_L = \frac{Ph}{Th} \quad (3)$$

where I_L is the life forms index, *Ph* is the frequency of Phanerophytes and *Th* is the frequency of Therophytes.

$$I_D = \frac{Sarco}{Sclero} \quad (4)$$

where I_D is the dispersal types index, *Sarco* is the frequency of Sarcochory and *Sclero* is the frequency of Sclerochory.

These indices calculated for each plot, compared the relative evolution of each pair of indicators between the different soil degradation classes. The higher the index, the greater the relative abundance of the biodiversity indicator in the numerator. The lower the index, the greater the relative abundance of biodiversity indicators at the denominator. Thereafter, the species richness (S), the chorological index (I_C), the life forms index (I_L) and the dispersal types index (I_D) were submitted to discriminant analysis and ANOVA using R software [32].

2.3 Assessing impacts of soil degradation on vegetation structure

The cover of each species was visually estimated within each plot. Braun Blanquet cover/abundance scale [33] was used: +: rare, less than 1% cover, 1: 1–5% cover, 2: 5–25% cover, 3: 25–50% cover, 4: 50–75% cover, and 5: 75–100% cover. The cover data of all inventoried species through the phytosociological surveys were grouped into an abundance matrix of 22 plots x 133 species and submitted to the Multi Response Permutation Procedures (MRPP). MRPP is a nonparametric procedure for testing the hypothesis of no difference between two or more groups of entities [34]. This procedure was used to pairwise compare the described soil degradation classes based on the cover data of their species lists. The analysis was computed with PC-ORD for Windows Version 5 [28].

3. Results

3.1 Impacts of soil degradation on phytodiversity

3.1.1 Floristic composition

Table 2 presents the pairwise comparison of soil degradation classes based on the index of similarity of Jaccard. On this basis, none of the soil degradation classes was similar to another. Given that the analysis was performed on the presence/absence matrix, we were able to conclude that all soil degradation classes were dissimilar, according to the floristic list. However, we noticed that the floristic composition of the vegetation in slightly and moderately degraded soils, although dissimilar, was closest (index of similarity of Jaccard equals to 0.434). Considering the discriminant species of each degradation class, the greatest number of discriminant species were found on slightly and moderately degraded soils (5 plants species) while the lowest were found on highly degraded soils (2 plant species) (**Table 3**).

3.1.2 Species richness, chorological types, life forms and dispersal types

The first two canonical axes obtained from the discriminant analysis on indicators of biodiversity were significant because they explained 97.59% of the initial information. The correlation between the two axes and the indicators of biodiversity showed that all the indicators (species richness, chorological, life forms and dispersal types indexes) were well and positively correlated with the first axis (0.91, 0.99, 0.99, 0.98 respectively) (**Table 4**). Thus, the first axis described high values of species richness and high values of chorological, life forms and dispersal type indexes. None of the indicators of biodiversity were well correlated with the second axis (**Table 4**).

The **Figure 2** showed that slightly and moderately degraded soils were positively correlated with the first axis while high and extreme degraded soils were well negatively correlated with the same axis. Based on the information gathered on this axis we could conclude that slightly and moderately degraded soil showed the highest species richness and were characterized by the highest relative abundance of regional species, phanerophytes and sarcochory. On the other hand, highly and extremely degraded soils showed lower species richness and highest relative abundance of species with wide distribution, therophytes

Soil classes	Light	Moderate	High	Extreme
Light	—	—	—	—
Moderate	0.434	—	—	—
High	0.149	0.19	—	—
Extreme	0.143	0.184	0.088	—

Table 2.
Index of similarity of Jaccard.

Species	Soil classes	Probability
Cochlospermum planchonii Hook.f.	Light	0.0062
Crossopteryx febrifuga (G. Don)Benth.	Light	0.0148
Indigofera nigritana Hook. f.	Light	0.0202
Strychnos spinosa Lam.	Light	0.0302
Hexalobus monopetalus (A.Rich.)Engl. & Diels	Light	0.0374
Basilicum polystachion (L.) Moench.	Moderate	0.0012
Blumea crispata Merxm. & Roessler var. cripata	Moderate	0.0012
Elephantopus mollis Kunth	Moderate	0.0032
Andropogon pseudapricus Stapf	Moderate	0.0230
Cissus corylifolia (Baker) Planch.	Moderate	0.0230
Stylosanthes fruticosa (Retz.)Alston	High	0.0010
Polygala multiflora Poir.	High	0.0084
Spermacoce filifolia (Schmach. & Thonn.) J.-P Lebrun & Stork	Extreme	0.0002
Cochlospermum tinctorium Perr. ex A.Rich	Extreme	0.0134
Chamaecrista mimosoides (L.) Greene	Extreme	0.0198

Table 3.
Discriminant species of each soil degradation class.

Variables	Can 1	Can 2
Species richness (S)	0.91	−0.34
Chorological index of disturbance (I_C)	0.99	0.02
Life forms index of disturbance (I_L)	0.99	0.14
Dispersal types index of disturbance (I_D)	0.98	0.13

Table 4.
Correlation between biodiversity indicators and the two canonical axes.

and sclerochory (or lower relative abundance of regional species, phanerophytes and sarcochory).

Simple statistics and ANOVA were summarized in **Table 5** and demonstrated that the between soil degradation classes based on biodiversity indicators were significant. Weighted spectrums of chorological types, life forms and dispersal types of diaspores were illustrated in **Figure 3(a–c)**. The highest species richness was found on slightly and moderately degraded soils (30.5 ± 7.2; 31.33 ± 4.93) and the lower values of this variable were found on highly (11.33 ± 3.21) and extremely degraded soils (16.5 ± 12.08). The high values of chorological index, life forms index and dispersal types index characterized light degraded soils (respectively 5.83 ± 1.64; 6.21 ± 3.82; 2.20 ± 0.76) and these values decreased gradually on moderately degraded soils (3.45 ± 0.40; 2.73 ± 1.70; 1.70 ± 0.91) and highly degraded soils (2.44 ± 0.096; 0.89 ± 1.54; 1.08 ± 0.38) and reached the lowest values on extreme degraded soils (1.51 ± 0.62; 0.78 ± 0.38; 0.62 ± 0.79). In other words, regional species, phanerophytes and sarcochory presented a regressive trend from light to extreme degraded soils through moderate and high soil degradation classes while species with wide distribution, therophytes and sclerochory followed a contrary trend.

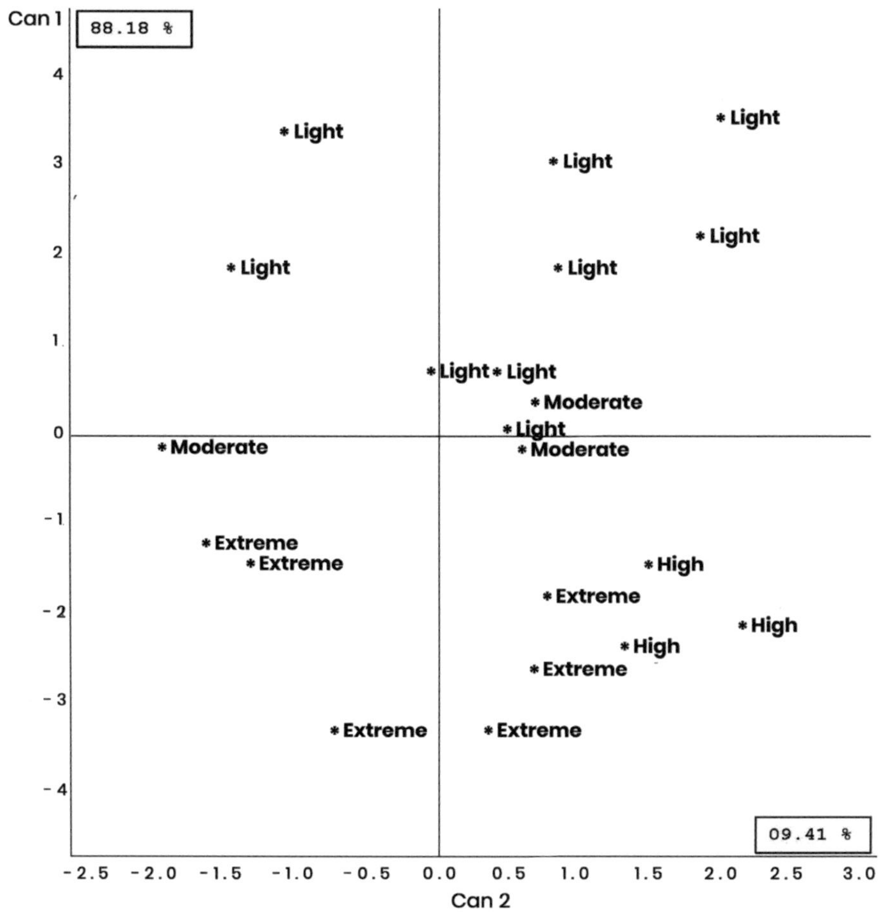

Figure 2.
Projection of soil degradation classes in the canonical system axis based on biodiversity indicators.

3.2 Impacts of soil degradation on vegetation structure

Tables 6 and 7 summarize the results of MRPP computed on cover data of each plots. First, all the degradation soil classes were considered together (**Table 6**). Thereafter, the degradation soil classes were considered two by two (**Table 7**). Considering all soil degradation classes, the results showed that the vegetation cover data for the four soil degradation classes were significantly different (**Tables 5** and **6**). However, the pairwise comparison (**Table 7**) gave more details and showed that the vegetation cover data of moderately and highly degraded soils were broadly overlapping ($p > 0.05$). Moderate and high degraded soils presented a relative similar vegetation type i.e. shrub savannas.

4. Discussion

4.1 Impacts of soil degradation on phytodiversity

The similarity index of Jaccard was significantly different on all the soil degradation classes and revealed that all soil degradation classes were dissimilar, depending on the floristic composition. The results allowed us to conclude that soil degradation induced modification of the floristic composition of vegetation. This finding

DOI: http://dx.doi.org/10.5772/intechopen.93899

Degradation degree	Light		Moderate		High		Extreme		F value	Pr (>F)
	m	s	m	s	m	s	m	s	—	—
Biodiversity indicators										
Species richness (S)	30.5	7.2	31.33	4.93	11.33	3.21	16.5	12.08	6.648	0.00324**
Chorological index (I_C)	5.83	1.64	3.45	0.40	2.44	0.096	1.51	0.62	17.73	1.31e-05***
Life forms index (I_L)	6.21	3.82	2.73	1.70	0.89	1.54	0.78	0.38	5.898	0.00549**
Dispersal types index (I_D)	2.20	0.76	1.70	0.91	1.08	0.38	0.62	0.79	5.956	0.00526**

m: mean; s: standard of deviation ns: not significant difference.
***Significant at 0.01.*
****Significant at 0.001.*

Table 5.
Mean, standard deviation and ANOVA of biodiversity indicators soil on each soil degradation classes.

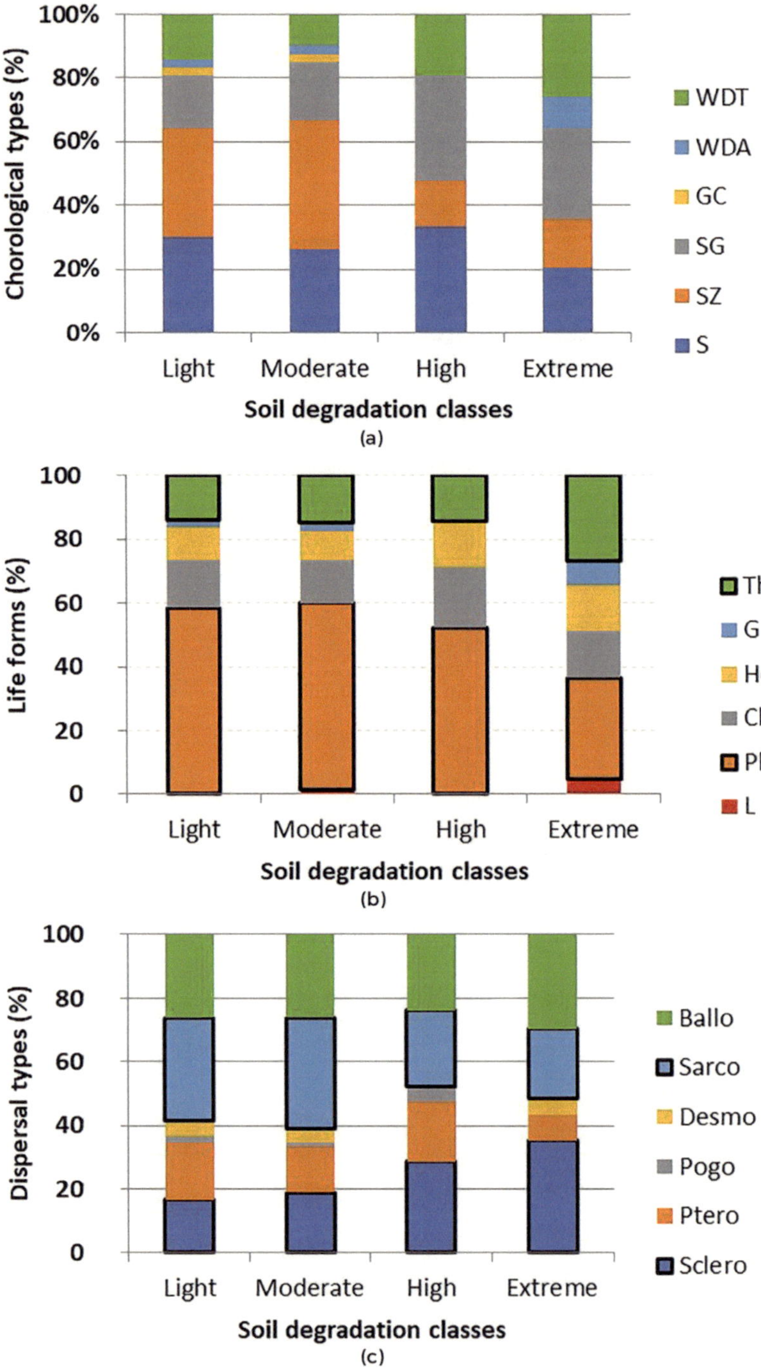

Figure 3.
(a) Weighted spectrum of chorological types, (b) life forms and (c) dispersal types on soil degradation classes. SG: Sudano-Guinean, SZ: Sudano-Zambezian, S: Sudanian/Th: Therophytes, G: Geophytes, Hc: Hemicryptophytes, Ch: Chamephytes, Ph: Phanerophytes, L: Lianas / Ballo: Ballochory, Sarco: Sarcochory, Desmo: Desmochory, pogo: Pogonochory, Ptero: Pterochory, Sclero: Sclerochory.

could be explained by the fact that the soil aggregate stability is closely related to soil organic matter composition [35], biological activity [36], infiltration capacity [37], water absorption and retention in the biomass and upper rhizosphere [38, 39] and erosion resistance [37]. Physical soil degradation on the hillsides of Atacora

Soil classes	Size	Average distance	A	T	P
Light	10	0.12	0.42486681	−6.7528582***	0.00000064
Moderate	3	0.14			
High	3	0.29			
Extreme	6	0.15			

A: Chance-corrected within-group agreement P: Probability of a smaller or equal delta T: Test statistic.
****Significant at 0.001.*

Table 6.
Global comparison with multi response permutation procedures.

Soil classes compared	A	T	P
High vs. Moderate	0.17460317	−1.68534137*	0.05423789
High vs. Light	0.21250178	−3.18865435**	0.00515585
High vs. Extreme	0.37185184	−3.98260352**	0.00360514
Moderate vs. Light	0.14571143	−2.20526374**	0.02559057
Moderate vs. Extreme	0.30172839	−3.95512754**	0.00298946
Light vs. Extreme	0.30594002	−6.68470594***	0.00005983

Chance-corrected within-group agreement P: Probability of a smaller or equal delta T: Test statistic.
**Significant at 0.1.*
***Significant at 0.05.*
****Significant at 0.001.*

Table 7.
Pairwise comparisons with multi response permutation procedures.

mountain was characterized by the removal of the organic layer and the modification of soil structure leading to the occurrence of ferricrete (extremely degraded soils) [17]. Soil degradation had resulted in soil loss, nutrient depletion, changes in soil structure, and soil hardening that limited plant root system penetration. Thus, only the most adapted species to the soil conditions were found on each soil degradation classes.

Moreover, the changes in species lists have been accompanied by a decrease of species richness and the number of regional species, phanerophytes and sarcochory as opposed to the number of species with wide distribution, therophytes and sclerochory. Many studies about post crop plant succession in Africa, United States and Europe [18, 21, 40, 41], or forest regeneration [7, 42] had shown that therophytes and sclerochory were pioneer species, which well-developed on disturbed areas, while phanerophytes and sarcochory colonized less disturbed areas. Moreover, according to references [23, 43, 44], therophytes and sclerochory developed a "ruderal" life strategy (habitat with high disturbance) and were submitted to a reproductive strategy of type r (rapid growth, effective dispersal and great invest in reproduction) while phanerophytes and sarcochory developed "competitive" or "stress-tolerant" strategies (habitat with low disturbance) and were submitted to a reproductive strategy type K (slow growth, effective use of resources and low invest in reproduction). The results then suggest that soil degradation leads to a loss of biodiversity and disturbance of vegetation.

As far as chorological types are concerned, we have reached the same conclusion of disturbance gradient. Indeed, regional species considered as indigenous or native species are found in great number in undisturbed areas and their number decrease along the gradient while species with wide distribution or immigrant species increase in number and are numerous on very disturbed areas [21, 45]. Thus, the vegetation trend over the different soil degradation classes followed a retrograde succession from the least disturbed soils (slightly degraded soils) to the most disturbed soils (extremely degraded soils) through intermediate stages (moderately and highly degraded soils).

4.2 Impacts of soil degradation on vegetation structure

Vegetation cover was used in the study as a measure of vegetation structure. With respect to vegetation cover data, the results showed that only moderately and highly degraded soil vegetation cover data were significantly similar ($p > 0.05$). Vegetation cover data provide information on vegetation type and may be used in gradients studies to investigate the effects of environmental factors on plant abundance [46, 47]. These results could be explained by the vegetation type found on each soil degradation class. Shrub savannas were the vegetation type found both on moderately and highly degraded soils. The types of vegetation observed on slightly and extremely degraded soils are tree/shrub savannas and herbaceous savannas respectively. The results of the impacts of soil degradation on vegetation structure namely vegetation type demonstrated the abundance of phanerophytes on slightly degraded soils, a decrease of the abundance of phanerophytes to the profit of therophytes on intermediate degradation classes and an abundance of therophytes on extremely degraded soils.

5. Conclusion

Soil degradation impacts vegetation in various ways. Floristic composition (presence/absence of species), species richness, chorological, life forms, dispersal types and vegetation type (tree and shrub savannas on light degraded soils, shrub savannas on high degraded soils and grass savannas on extreme degraded soils) were the different aspects of vegetation which were modified along the gradient of soil degradation. The overall trend observed, showed the degradation of vegetation along the gradient of degradation of soils. The findings confirmed the negative impact of land degradation on vegetation and plant diversity. The results provided a good overview of the relationship between soil degradation and vegetation, useful for management policies. The study did not attempt to characterize the vegetation found on each degradation class, but rather to test the effects of soil degradation gradients on some measures of phytodiversity and vegetation structure. However, one limitation of this evaluation could be the low number of plots considered, which makes it difficult to generalize the results at the level of the whole study area. Further researches should be conducted in order to eliminate the limitation.

Acknowledgements

This work was entirely supported by UNDESERT project (EU FP7 243906), "Understanding and combating desertification to mitigate its impact on ecosystem services" funded by the European Commission, Directorate General for Research and Innovation, Environment Programme for financial support. The main goal of the project was to raise the understanding of how degradation and desertification processes affect biodiversity, soil and human livelihoods.

Author details

Farris Okou*, Achille Assogbadjo and Brice Augustin Sinsin
Laboratory of Applied Ecology, University of Abomey-Calavi, Abomey-Calavi, Benin

*Address all correspondence to: farrisokou@gmail.com

References

[1] Bao Le Q, Nkonya, E, Mirzabaev, A. Biomass productivity-based mapping of global land degradation hotspots. ZEF Discussion Papers on Development Policy No.193, Bonn: Zentrum für Entwicklungsforschung (ZEF). 2014.

[2] Le Q B, Tamene L, Vlek P L G. Multi-pronged assessment of land degradation in West Africa to assess the importance of atmospheric fertilization in masking the processes involved. Global Planet. Change Global and Planetary Change. 2012. 92-9371-81. DOI: http://dx.doi.org/10.1016/j.gloplacha.2012.05.003.

[3] Anderson P M, Hoffman M T. The impacts of sustained heavy grazing on plant diversity and composition in lowland and upland habitats across the Kamiesberg mountain range in the succulent Karoo, South Africa. Journal Of Arid Environments. 2007;70(4), 686:700. DOI: http://dx.doi.org/10.1016/j.jaridenv.2006.05.017.

[4] Wang X, Dong S, Sherman R, Liu Q, Liu S, L, Y, Wu Y. A comparison of biodiversity-ecosystem function relationships in alpine grasslands across a degradation gradient on the Qinghai-Tibetan Plateau. The Rangeland Journal. 2014;37(1), 45:55. DOI: http://dx.doi.org/10.1071/RJ14081.

[5] Andrade B O, Koch C, Boldrini I I, Vélez-Martin E, Hasenack H, Hermann J-M, Kollmann J, Pillar VD, Overbeck G E. Grassland degradation and restoration: a conceptual framework of stages and thresholds illustrated by southern Brazilian grasslands. NCON Natureza & Conservação. 2015;13(2), 95:104. DOI: http://dx.doi.org/10.1016/j.ncon.2015.08.002.

[6] Ruiz-Jaén, M C, Aide T M. Vegetation structure, species diversity, and ecosystem processes as measures of restoration success. Forest Ecology and Management Forest Ecology and Management. 2005;218(1-3), 159:173. DOI: http://dx.doi.org/10.1016/j.foreco.2005.07.008.

[7] Vroh B T A, Adou Yao C Y. Successional dynamics of tree species during forest recovery in the southeast of Côte d'Ivoire. Sciences de la vie, de la terre et agronomie - CAMES. 2017;5(2), 30:38.

[8] Padonou E A, Assogbadjo A E, Sinsin B, Bachmann Y. How far bowalization affects phytodiversity, life forms and plant morphology in Sub-humid tropic in West Africa. Afr. J. Ecol. African Journal of Ecology. 2013;51(2), 255:262.

[9] Parras-Alcantara L, Martin-Carrillo M, Lozano-Garcia B. Impacts of land use change in soil carbon and nitrogen in a Mediterranean agricultural area (Southern Spain). Solid Earth Solid Earth. 2013;4(1), 167:177. DOI: http://dx.doi.org/10.5194/se-4-167-2013.

[10] Keesstra S D, Maroulis J, Argaman E, Voogt, A, Wittenberg, L. Effects of controlled fire on hydrology and erosion under simulated rainfall. Cuadernos de Investigación Geográfica. 2014;40(2), 269. DOI: http://dx.doi.org/10.18172/cig.2532.

[11] Brevik E C, Cerda A, Mataix-Solera J, Pereg L, Quinton J N, Six J, Van Oost K. The interdisciplinary nature of Soil. 2015;1(1), 117:129. DOI: http://dx.doi.org/10.5194/soil-1-117-2015.

[12] Adegbidi A, Gandonou E, Mulder I, Burger K,. Farmer's perceptions and sustainable land use in the Atacora, Benin. Creed Working Paper. 1999;22, 1:51.

[13] Mulder I. Soil Degradation in Benin: farmers' Perceptions and Responses

[PhD Thesis]. Amsterdam: Thela Thesis; 2000

[14] Tente B. Dynamique actuelle de l'état de surface dans le massif de l'Atacora : Secteur Perma-Toucountouna [Master thesis]. Abomey-Calavi : FLASH, Université Nationale du Bénin; 2000.

[15] Saïdou A, Kuyper T W, Kossou D, Tossou R C, Richards P. Sustainable soil fertility management in Benin: learning from farmers. NJAS wageningen journal of life sciences. 2004;52(3/4), 349:369. DOI : http://dx.doi.org/10.1016/S1573-5214(04)80021-6.

[16] Tente B. Recherche sur les facteurs de la diversité floristique des versants du massif de l'Atacora: Secteur Perma-Toucountouna (Bénin) [PhD thesis]. Abomey-Calavi : Ecole Pluridisciplinaire "Espaces, Cultures et Développement", Université d'Abomey-Calavi; 2005.

[17] Okou F A Y, Assogbadjo A E, Bachmann Y, Sinsin B,. Ecological Factors Influencing Physical Soil Degradation in the Atacora Mountain Chain in Benin, West Africa. Mountain Research and Development. 2014;34(2), 157:166. DOI:10.1659/MRD-JOURNAL-D-13-00030.1.

[18] Debussche M, Escarré J, Lepart J, Houssard C, Lavorel S. Changes in Mediterranean plant succession: Old-fields revisited. Journal of Vegetation Science. 1996;7(4):519:526. DOI: 10.2307/3236300

[19] Sieglstetter R, Wittig R. L'utilisation des ligneux sauvages et son effet sur la végétation de la région d'Atakora (Bénin nord-occidental). Etudes flor. Vég. Burkina Faso. 2002;723:30.

[20] Wala K. La végétation de la chaîne de l'Atacora au Benin : Diversité floristique, phytosociologique et impact humain [PhD thesis]. Lomé, Togo : Faculté des Sciences, Université de Lomé Thèse de Doctorat; 2005.

[21] Bangirimana F, Masharabu T, Bigendako MJ, Lejoly J, de Cannière C, Bogaert J. Evolution des paramètres floristiques au cours de la dynamique post culturale dans les jachères du site Bibara dans le Parc National de la Ruvubu (Burundi). Bulletin Scientifique de l'Institut National pour l'Environnement et la Conservation de la Nature (INECN). 2009;7(3):13

[22] Shrestha R P. Land degradation and biodiversity loss in Southeast Asia. In: Alkemade R, Shrestha R P, Trisurat Y, editors. Land Use, Climate Change and Biodiversity Modeling: Perspectives and Applications. IGI Global; 2011. p. 303-327.

[23] Grime JP, Pierce S. The Evolutionary Strategies That Shape Ecosystems. John Wiley & Sons; 2012

[24] Weber H E, Moravec J, Theurillat J P. International Code of Phytosociological Nomenclature. 3rd edition. JVS Journal of Vegetation Science. 2000;11(5), 739:768.

[25] R Core Team. R: A language and environment for statistical computing. R Foundation for Statistical Computing, R: A language and environment for statistical computing; 2017. https://www.R-project.org/ https://www.R-project.org/.

[26] Mucina L, Maarel Van Der E. Twenty years of numerical syntaxonomy. Vegetatio. 1989;**81**(1-2):1:15. DOI: 10.1007/bf00045509

[27] Dufrene M, Legendre P. Species Assemblages and Indicator Species: The Need for a Flexible Asymmetrical Approach. Ecological Monographs. 1997;67(3), 345:366. DOI:10.2307/2963459.

[28] McCune B, Mefford M J. PC-ORD : Multivariate analysis of ecological data. MjM software design, Gleneden Beach, Oregon, USA; 1999.

[29] Djego, J G M. Phytosociologie de la végétation de sous-bois et impact écologique des plantations forestières sur la diversité floristique au sud et au centre du Bénin. [PhD thesis]. Abomey-Calavi : Ecole Pluridisciplinaire "Espaces, Cultures et Développement", Université d'Abomey-Calavi; 2006.

[30] Pidwirny M, Jones S. Introduction to the Biosphere-Plants Succession, Fundamentals of Physical Geography. 2nd ed. Columbia, Okanagan: University of British; 2010

[31] Adomou AC. Vegetation patterns and environmental gradients in Benin: Implications for biogeography and conservation. [PhD thesis]. In: Wageningen : Wageningen University. 2005

[32] R Core Team. R: A language and environment for statistical computing. R Foundation for Statistical Computing, R: A language and environment for statistical computing; 2019. https://www.R-project.org/ https://www.R-project.org/.

[33] Westhoff V, Van Der Maarel E. The Braun-Blanquet Approach. Springer. 1978;287:399. DOI : http://dx.doi.org/10.1007/978-94-009-9183-5_9.

[34] McCune B, Grace J B. Analysis of ecological communities. MjM Software Design; 2002.

[35] Tisdall J M. Formation of soil aggregates and accumulation of soil organic matter. In: Carter M R, Stewart B A, editors. Structure and organic matter storage in agricultural soils. CRC Press: Boca Raton, FL, USA; 1996. p. 57-96.

[36] Wander M, Traina S, Stinner B, Peters S. Organic and conventional management effects on biologically active soil organic matter pools. Soil Science Society of America Journal. 1994;58(4), 1130:1139. DOI : http://dx.doi.org/10.2136/sssaj1994.03615995005800040018x.

[37] Pierson F B, Blackburn W H, Vactor S, Wood, J. Partitioning small scale spatial variability of runoff and erosion on sagebrush rangeland. JAWRA Journal of the American Water Resources Association. 1994;30(6), 1081:1089. DOI : http://dx.doi.org/10.1111/j.1752-1688.1994.tb03354.x.

[38] Johnston M H. Soil-vegetation relationships in a Tabonuco Forest Community in the Luquillo Mountains of Puerto Rico. Journal of Tropical Ecology. 1992;8(3), 253:263. DOI : http://dx.doi.org/10.1017/S0266467400006477.

[39] Chen ZS, Hsieh CF, Jiang FY, Hsieh TH, Sun IF. Relations of soil properties to topography and vegetation in a subtropical rain forest in southern Taiwan. Plant Ecology. 1997;**229**:241

[40] McCook L J. Understanding ecological community succession: Causal models and theories, a review. Vegetatio 1994;110(2), 115:147. DOI:10.1007/bf00033394. DOI : http://dx.doi.org/10.1007/BF00033394.

[41] Smit R. The Colonization of Woody Species in Old Fields: Old Field Succession in the Netherlands. Agricultural University Wageningen; 1996.

[42] Koubouana F, Ngoliele A, Nsongola G. Evolution des paramètres floristiques pendant la régénération des forêts de la réserve de la Lefini (Congo brazzaville). Annales de l'Université Marien NGOUABI. 2007;8(4), 10:21

[43] MacArthur R H, Wilson O. The theory of island biogeography. Princeton, NJ; 1967. DOI : http://dx.doi.org/10.1515/9781400881376.

[44] Grime J P. Evidence for the existence of three primary strategies

in plants and its relevance to ecological and evolutionary theory. The American Naturalist. 1977;111(982), 1169:1194. DOI : http://dx.doi.org/10.1086/283244.

[45] Domenech R, Vila M. The role of successional stage, vegetation type and soil disturbance in the invasion of the alien grass *Cortaderia selloana*. Journal of Vegetation Science. 2006;17(5), 591:598. DOI:10.1658/1100-9233(2006)17[591:trossv]2.0.co;2. DOI : http://dx.doi.org/10.1111/j.1654-1103.2006.tb02483.x.

[46] Austin M. Species distribution models and ecological theory: A critical assessment and some possible new approaches. Ecological Modelling 2007;200(1-2), 1:19. DOI : http://dx.doi.org/10.1016/j.ecolmodel.2006.07.005.

[47] Damgaard C. On the distribution of plant abundance data. Ecological Informatics. 2009;4(2), 76:82. DOI : http://dx.doi.org/10.1016/j.ecoinf.2009.02.002.

Section 2

Experiences on Soil Conservation

Chapter 6

Erosion Control Success Stories and Challenges in the Context of Sustainable Landscape Management, Rwanda Experience

Jules Rutebuka

Abstract

The Government of Rwanda sets up a conducive policy environment to invest in several development initiatives. Agriculture sector as the main contributor in the economic development received supports to sustainably manage Rwandan hilly landscape, dominantly ranging from 5 to 55% slope gradient. Intensive erosion control interventions confronted with different approaches have been introduced in the country such as participatory landscape management, (participatory) integrated watershed management and site-located intervention without any specified approach. This chapter intends to describe and evaluate the impacts of these previous approaches used in Rwanda in order to retrieve the success stories and encountered challenges as lessons learnt in the future interventions for optimizing land productivity in a sustainable manner. Participatory landscape approach in Gishwati area was a success story in protecting degraded lands and generating ecosystem benefits. It leads to more sustainable natural resources management from participatory planning up to implementation which addressed the frequent landslides, erosion and flooding while sustainably exploit the land to the profit of local farmers in the livelihoods. About 6,600 ha of lands have been successfully protected with full-packaged bench terraces, rangeland blocks and forest regeneration. This participatory approach also helped to relocate people from high risk zones to other safe places and build capacities of farmers through farm-livestock cooperatives. On the other side, Nyanza and Karongi sites under LWH project also emphasized strong evidences how land husbandry technologies (terraces) efficiently reduced erosion risks and improved farmers' livelihoods. Lands were made productive with implementation of bench terraces on 3212 and 2673 hectares respectively for the two selected sites. However, challenges were observed from technical and socio-economic contexts which might have caused farmers to abandon or under-exploit the terraced lands. Finally, the chapter suggests to scale up the participatory landscape management approach which supports the involvement of farmers' communities in the process.

Keywords: erosion, terraces, successes, challenges, participatory, landscape, Rwanda

1. Introduction

Rwanda, the country of thousand hills, has a small coverage area of 26,338 km^2 with the highest (rural) population of 12 million inhabitants (416 habitants per km^2), among African countries. More than 80% of population depends on agriculture sector which is dominated by subsistence farming at average farm size of 0.5 hectare [1].

Over the last two decades, the Government has experienced tremendous and steady rates of economic growth nationwide averaging 5.7% in 2019 [2]. While this sector contributes approximately to about 27% of the national GDP and 68% of the labor force [1], there is an intense pressure on degradation of natural resources especially land and water, by occupying marginal and non-protected lands. Thus, agriculture is still affected by low productivity due to several factors. Among others, Rwandan biophysical environment is dominantly characterized by steep slopes accentuated from Eastern to Western facings, and this mountainous topography exposes soil to water erosion risks, especially in the Highlands of Nothern and Western parts of Rwanda. Particularly, erosion risk is chiefly associated with slope ranges from 5 to 55% on arable land (about 48% of the total area) [3–5].

The combination of soil erosion, climate change condition, poor soil fertility and inappropriate steepland managements have aggravated such low productivity levels. In addition, intensive farming activities resulted into pollution, lowland siltation, soil nutrient depletion and soil acidity [6–8]. The acidic soils cover about 50% of national land area [9, 10]. Recently, climate change conditions have also to harmonize style and droughts reduced the performance of agriculture production system, resulting from to rainfall differences as affected by El Niño - Southern Oscillation (ENSO) events (El Niño and La Niña) [11–13]. This renders small-scale, subsistence, rain-fed farming vulnerable and leads to more advanced land degradation problems.

In the framework of finding appropriate solutions to combat land degradation problems, the country sets up a conducive environment with strategic policy tools since the past 20 years such as Vision 2020, Strategic Plan for Agriculture Transformation (PSTA I, II, III, IV). Recently, National Strategy for Agriculture Transformation (NST1) (2017–2024) and its Forth Strategic Plan for Agriculture and Transformation (PSTA4, 2018–2024) identified increasing productivity and resilience through sustainable land management approach as one of the priority areas in the economic development. Different actions from policy and development aspects had been invested in soil erosion control systems using a wide range of erosion control measures chiefly terraces, towards sustainable environment protection and agricultural transformation pathways. Intensive erosion control interventions confronted with different approaches bringing both on-site and off-site impacts [14, 15]. They adopted either different ways such as participatory landscape management or (Participatory) integrated watershed management. Thirdly, none of them was adopted to establish soil erosion control techniques.

Therefore, this chapter intends to describe and evaluate the impacts of different approaches used in erosion control systems in Rwanda in order to retrieve the success lessons, but also pinpoint challenges of each approach used. The chapter is practically assessing land husbandry interventions undertaken in two government projects namely Gishwati Water and Land Management (GWLM), and Land, Water-harvesting and Hillside-irrigation (LWH) for gaining understandings of the success or issues to be considered in the future interventions in the country as well as in other areas with similar landscape conditions. From lesson learnt, the chapter intended to recommend the best and comprehensive technical strategies aligning to land husbandry in rural farming systems for improving sustainable landscape management and optimize land's productivity.

2. Approach

This chapter compiles two case studies in Rwanda describing and analyzing various approaches used on soil erosion control systems in the recent past years (after 2000). The findings have to inform the success experiences, problems encountered and generated potential policy and technical recommendations to be adopted in future.

The first section concerns the north-western part of Rwanda namely Gishwati area using participatory landscape approach. The second involves the use of two watersheds (Nyanza-23 and Karongi 12–13) developed under support of World bank project namely Land, Water-harvesting and Hillside-irrigation (LWH) which adopted integrated watershed management approach. The last section discussed lessons learnt to inform policy decision makers at national, regional, international scopes of what is the appropriate way to sustainably optimize the land productivity based on Rwandan experience. The study areas are located from different agro-ecological zones: Gishwati site in the Birunga, Nyanza in central plateau, and Karongi in Kivu lake Borders (**Figure 1**).

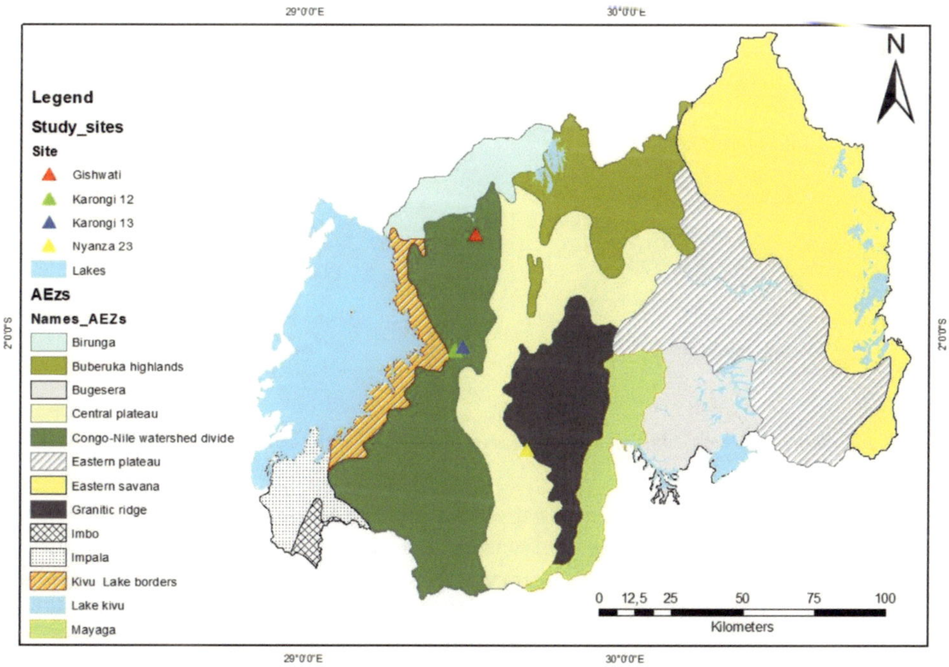

Figure 1.
Map of Rwanda showing its agro-ecological zones and study sites (Nyanza 23, Gishwati, and Karongi 12–13). (Source: Author).

3. Sustainable landscape management approach: Gishwati case

3.1 Description of the study area

Gishwati targeted area covers 6,600 ha across Jenda, Karago, Rambura and Bigogwe sectors of Nyabihu district, and Kanama, Nyakiliba and Kanzenze sectors of Rubavu district. This area constitutes 26.5% of the total Gishwati ecosystem in its northern part. The area is geographically located at latitude of 1.689418°S and longitude of 29.532433° E. The altitude varies from 2,191 to 2,959 m. Gishwati area had greatly suffered with problems of soil erosion, landslide, gully, flooding, human loss and destruction of development infrastructures after 1994 due to occupation of fragile forest reserve by mass return of refugees. Gishwati was before a natural forest ecosystem which has been converted to agriculture, livestock and settlement lands. Land was intensively exploited mainly for agriculture purposes such as cropping of Irish potatoes, climbing beans, peas, wheat, tea, etc., but also with livestock activities on scattered pasture grasses and poorly managed woodlots (**Figure 2**).

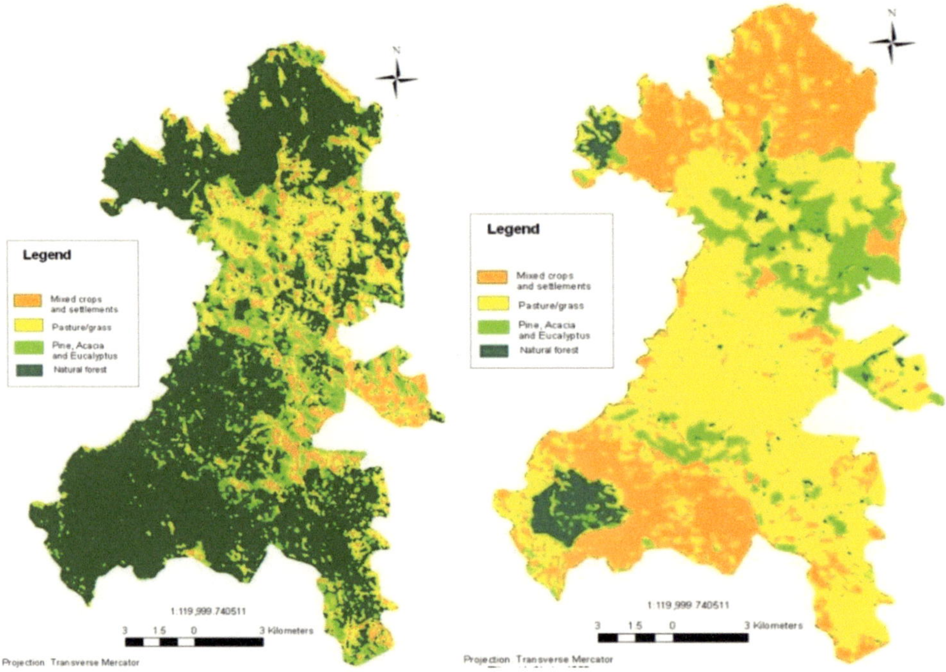

Figure 2.
Map of land cover changes of Gishwati ecosystem from 1986 to 2006 (left to right), Source: GWLM project.

As many places in the country, Gishwati is characterized by a complex lithology and landscape diversity due to elevation differences from valley bottom to mountain summits. Soils in Rwanda vary across very short distances due to the complexity of relief and parent materials [16, 17]. This observation varies from hill to hill and hilltop to the lower slope and valley bottom [17]. Any intervention for its success should consider this biophysical complexity.

The drastic change in land use affected local communities to live regularly with risks of landslide and floodings. These risks are subjected to high rainfall rainfall ranging from 1800 to 2500 mm per year and to fragility of soils (*Ruseseka* in Kinyarwanda local language) from forest soils and volcanic materials lying on a bed-rock at very steep slopes. All these factors together with inappropriate agriculture practices and lack of land and water management measures induced very severe erosion. **Figure 3** demonstrates how eroded soil materials flooded the lowland (left) and leaving plantation tree outcropped (right).

3.2 Approach

Since 2010, Gishwati Water and Land Management Project (GWLM) has been initiated to effectively counteract the landslides, floods and erosion risks but also

DOI: http://dx.doi.org/10.5772/intechopen.96267

Figure 3.
Induced natural hazards: Landslide, erosion, flooding, silting, and root outcropped: Bigogwe, April 2007 (left) and April 2010 (right).

strengthening the potential for agriculture development in Gishwati area in the context of improving livelihood communities. The GWLM project of the Ministry of Agriculture and Animal Resources (MINAGRI) understood the vitality to sustainable restoring the landscape potentials of Gishwati and hence it developed an approach which consists of two resolutions:

- Harmonizing the healthy co-existence of the agrarian communities with the fragile ecosystem of Gishwati;

- Maximizing sustainable economic contribution of Gishwati to the communities' improved way of life.

In this context, the MINAGRI concerted efforts of technical/scientific expertise from potential actors including the local government, the beneficiary farmers, different Government institutions such as MINAGRI, Ministry of Environment, Lands and Water (currently MOE-Ministry of Environment), and other relevant agencies/organizations to support the project goal. This aligns with participatory landscape approach by which key stakeholders contributing to economic development should intervene to establish a comprehensive approach for harmonizing the healthy co-existence of the agrarian communities with the fragile ecosystem of Gishwati.

The core issue of the Gishwati was to find a best way by which land degradation issues would be successfully avoided by linking different soil and water management interventions to the different land potential units of the project area while supporting sustainable existence between human needs and natural resource-based opportunities. A participatory and integrated landscape approachis considered to improved management of natural resources to support sustainable agricultural productivity but also taking into account the effects of climate change. Adoption of landscape approach puts attention to modernizing land and water management technologies as well as promoting extension services that effectively guaranty stability of sloping lands within Wet Rainfall Regimes of Gishwati.

The government realized that the intervention is of momentous challenge to assure stable and resilient environment using scientific-based technologies. Another aspect considered how the fertile Gishwati soil and the year-round rainfall contributes to the improved livelihood of the communities. This calls upon using ingeniously designed physical and biological technologies that guaranty the sustainability and productivity of the land through effective water and soil management practices. To materialize the economic potential of land husbandry technologies, farmers are encouraged to consolidate their lands for construction of suitable and long-stretching land husbandry structures which guaranteed the increased and continued production of crop value chains of the project area in Nyabihu and Rubavu districts. The landscape restoration of this area has been supported by the policies, among others, the land consolidation, the crop intensification, transformation of subsistence to market-based agriculture, etc., as set by the MINAGRI.

3.2.1 Criteria for selecting appropriate land uses and managements

Factors were itemized in order to define every land unit according to its potentials. Pratically, it concerned placement of different soil and water management interventions on the appropriate land potential units of the project area. The following criteria were considered:

- Consulting and exploit existing datasets for Rwanda, specifically in the concerned region;
- Understanding the nature of the slope gradient;
- Exploiting the soil depth and characteristics of the project area.

3.2.1.1 Exploiting available datasets

The agro-climatic data of the Gishwati area have been gathered for analysis of rainfall variability and agressivity. This area shows agroclimatic zones of wet highland, wet frost and wet alpine frost conditions. The information of wet moisture regimes with very limited evapotranspiration in high altitudes could be considered in the equation for generating appropriate interventions. Soil database was also explored to understand the soil properties including soil depth, and soil types.

3.2.1.2 Consideration of slope gradient nature

To understand the impact of topography, the map of slope gradient of Gishwati watershed using ArcGIS spatial analysis tools has been generated from the Digital Elevation Model (DEM −30 m resolution), accessed from the United States Geological Survey (USGS) database (http://earthexplorer.usgs.gov/). Digital representation topography (DTM) generated from DEM helped to calculate five slope classes (0–6, 6–16, 16–40, 40–60 and > 60%) (**Figure 4**).

Outputs of the slope map generated provided distribution of slope classes as follow: 50% of the Gishwati area (3290 ha) are located within 16–40%, 23% (1491 ha) within 40–60%, 13% (895 ha) within 6–16%, 10% (659 ha) within 0–6%, and finally 4% (279 ha) is above 60%. However, the forms of slope are so complex so that the slope criterion was not easy for defining recommendation zones [17]. Thus, scientists managed to agree on the approach for protecting this complex biophysical environment. RUSLE model helped to develop erosion risk assessment whereby slope factor contributed more (**Figure 4**).

3.2.1.3 Exploiting soil depth

Soil depth was assessed to see the storage medium for the year–round rainfall that could cause landslides as well as to understand the rooting depth required for the crops to be grown in the area. Field survey of soil depth identified three levels (0–50, 50–100 and 100–200+ cm) for each soil types within different slope categories of the study area using augering method. The pedological prospections were conducted

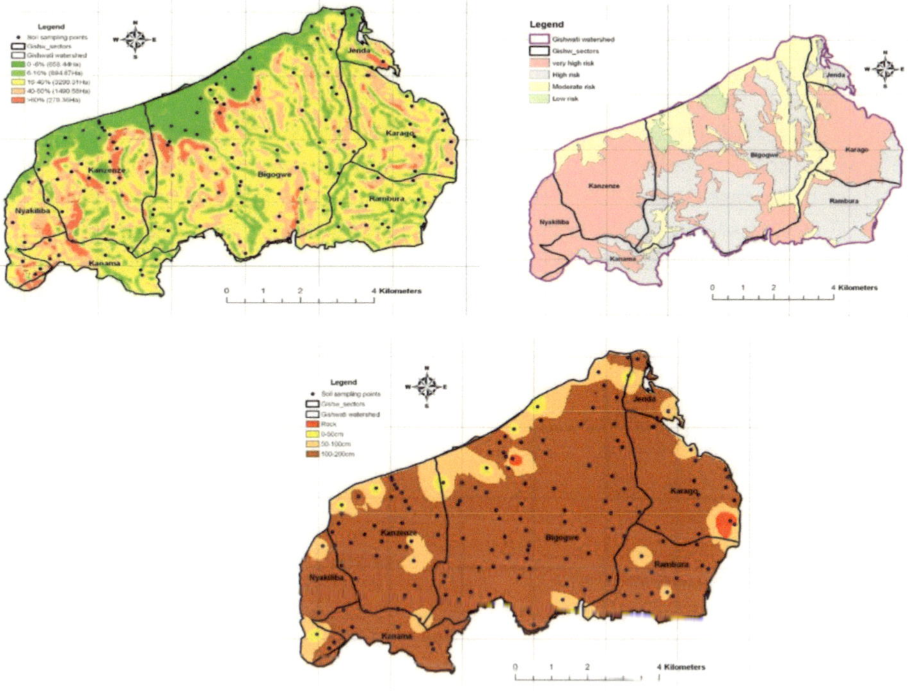

Figure 4.
Maps of slope gradient, soil depth, erosion risk and dominant soil types in Gishwati area. Source: GWLM project.

on 52 depth tests on the dominant slope class of 16–40%, 29 tests on 40–60% slope class, 34 tests on 6–16% slope class, 27 depth tests on 0–6% slope class, and 14 depth tests on >60% slope category (**Figure 4**). Based on the soil database of Rwanda, field pedological prospection resulted in 156 soil depth tests for dominant soil types such as *Andosols*, *phaeozems*, *Acrisols*, *Cambisols*, *Lixisols* and *Leptosols* [18].

Results showed that most soils are very deep and well developed. More than 80% of conducted auger tests ranged between 100–200+ cm depth. Decisions were taken accordingly to guide recommended options for restoring landscape. The soil depth discloses how soil material with water infiltration storage can exerts pressure over the sloping land. In addition, it also guides to know the relatively most appropriate crops to be grown over each type of soil.

Very shallow soil depth zones such as the bare rook-covered lands are recommended for area closure. The next shallow depth lands (depth of 0–50 cm) are recommended for the shallow rooting grasses (range land). The utilization of soils with depth between 50–100 cm and 100–200+ was variable regarding the combination with other bio-physical factors. If the same land category was in the moist to dry rainfall regime, one could easily recommend the 100–200+ depth land for trees (deep rooting) and the other for the relatively shallow rooting shrubs. On the other hand, the deep rooting on very steep slopes would encourage excessing waterload on the mass of the deep soil materials, hence landslide occurs. In this case, planting shrubs/trees is recommended.

3.2.1.4 Soil characteristics

Soils of Gishwati are dominantly underlaid on a bed rock. Shallow soils (0–50 cm depth) are mainl *Leptosols* and *Andosol*s, derived from recent volcanic ejecta along the Bigogwe plain. These soils are very fragile and less structured on steep slopes, hence prone to landlside as typical forest soils. They are called *Ruseseka* in Kinyarwanda by local farmers. *Andosols, Leptosols, and Phaezems* are the dominant soil types in Gishwati.

Regarding fertility potential, Rwanda digital soil dataset revealed that the fertilty of the soil is excellent for crop production in such year-round rainfall regime [19, 20]. In this context, soil analysis was done with top-soil samples collected between 0–30 cm soil depth layer for assessing soil nutrient and acidity status. It showed that Gishwati area has a great potential for agriculture in Rwanda once erosion is controlled. Nutrients were above the critical level except for phosphorus. Soil acidity problem in the area was quite low with soil pH range between 5.5 and 6.6 unlike to other Rwanda soils in the North-Western parts. The soils have high organic carbon content (3.2–5.1%) and significantly high contents of crop nutrients.

The study also recognized the main soil types including *Andosols, Phaeozems, Acrisols and Lixisols* on hillsides; *Leptosols* developed on recent volcanic materials along Musanze-Rubavu national road and *Cambisols* derived from colluvial and alluvial sediments in the narrow valleys. Andisols are developed in mild weathering conditions from volcanic eject while *Phaeozems* are developed in moist conditions under grassland or forest with a mollic epipedon. *Acrisols* are developed in wet tropical or subtropical forests, with acid silicates clays, iron and aluminum oxides. *Lixisols* were developed under moist or mildly acid conditions with acid clays accumulation (called "inombe" in Kinyarwanda); and *Leptosols* referred to as younger or recent soils derived from metamorphic parent materials. Specifically, *Andosols, Phaeozems, Leptosols and Cambisols* are very fertile and suitable for a wide range of crops, namely, Irish potatoes, maize, beans, peas, wheat, variety vegetables, etc.

3.2.1.5 Community-based factor

One of the key partners in the success of the project was the great involvement of the local community in the entire process. Participatory landscape approach considered the active participation of rural communities in order to invigorate people-centered solutions in the community livelihoods. Tantoh et al. [21] stated that promotion of rural resources can only be successful if rural communities are integrated and engaged in the land husbandry interventions. It helped increasing ownership of beneficiaries, even after. Leaders of farmers received training for participatory community land use plan and map that was translated from English to Kinyarwanda local language.

The restoration of Gishwati area used this participatory approach through Labour Intensive Public Works namely as HIMO (*Haute Intensité de Main d'œuvre or High Intensive Labor)*. The latter consists of using people in respective to their social classes towards enforcement of local beneficiaries for job creation purpose and availing income-generating activities. Local leaders, opinion leaders, farmer promoters and other farmers' organizations located in the area as well as national institutions were actively involved in the whole process of planning, relocation of population from land degradation risk zones up to the implementation of sustainable landscape-related solutions (**Figure 5**).

3.3 Result as success stories in sustainable landscape management

The above discussed factors for determining optimal and appropriate landscape management approaches were combined. To this effect, the guidance relied on soil depth, soil types and slope gradient as well as climatic information. Besides, proper

Figure 5.
Community involvement in planning and implementation of land and water management. Source: GWLM project.

community mobilization, and sensitization in the whole process of landscape restoration were critically important to the sustainable establish land husbandry interventions.

3.3.1 Participatory planning and implementation of land husbandry interventions

3.3.1.1 Involvement of stakeholders in promoting land husbandry technologies

1. Capacity and knowledge about sustainable landscape management have been expanded through trainings. Trainings were intensively conducted to increase knowledge and understanding of beneficiaries about sustainable and new improved land management techniques. It comprised also how farmers should sustain implemented land husbandry interventions.

2. Farmers were sensitized to be involved in the whole process of landscape restoration of Gishwati area. Activities of sensitization and mobilization have been conducted since project start up in 2010. The project beneficiaries played a big role in mass mobilization campaigns, meetings at all levels (villages, cells, sectors, districts and central government levels). Beneficiaries explored the problems of erosion, flooding, and landslides as well as their causes. They also provided possible suggestions about landscape restoration.

3. After trainings, about 13,056 beneficiaries were involved in and earned income from land husbandry works through locally created companies within HIMO approach.

4. A new pyrethrum crop has been established in the Gishwati area. Farmers benefited as well as the promotion of pyrethrum production through support of seedling provisions, trainings, field visits and other technical assistance (**Figure 6**).

5. For the sustainability of achieved project interventions, cooperatives were formed mainly aligning to pyrethrum, and Irish potato crop commodities in order to optimize production on developed land husbandry infrastructures. Cooperatives have been registered and certificates were issued by Rwanda Co-operative Agency (RCA). In addition, the project created 42 self-help groups (around 600) in Gishwati area for the development and management of land husbandry technologies and other ecosystem services.

Figure 6.
Field visits (left) and plantation (right) on pyrethrum grown in Gishwati area. Source: GWLM project.

3.3.1.2 Participation of beneficiairies in land redistribution in Gishwati area

Land redistribution was one of the challenging issues in Gishwati area to avoid any dispute of land among land users. After restoring landscape, all developed lands intednded for crop and grazing activities were redistributed back to local people. In this process, a technical team was formed including local leaders at district, sector, cell and village levels. Integrating beneficiaries (farmers) in the decision process of land use planning helped to ensure the sustainable utilization of implemented land husbandry technologies.

This activity was successfully implemented for 5633 households whereby lands were equitably allocated to 4353 and 1280 farmers for crop and grazing activities respectively. For the rangeland, each household was given one hectare. However, to ensure sustainable utilization and management of this land, households were formed into groups of ten, making a total area of ten hectares which were cut into one paddock.

3.3.2 Landscape restoration interventions in Gishwati area

In addition to identified land use plan categories, this section comprised implementation of land husbandry technologies in crop land, development of rangeland, plantation of forests, construction of road and water drainage infrastructures and other ecosystem products.

3.3.2.1 Identification of land sensitivity levels or resilient categories

The first and basic outcome of the project was the identification of different sensitivity levels in Gishwati area. Results pointed out the effective use of graded land management technologies based on the assessment of above discussed factors (slope, soil type, soil depth, and rainfall). Biological measures such as live-fences have been used to compartmentalize into blocks (**Figure 7**).

Twenty (20) land sensitivity level/resilient categories were identified referring to land use units and considered for the specific land and water management technologies. As shown in the **Table 1**, land units 1 and 2 at 603.8 ha (9.5% of the total area) were used for minor agriculture intervention using graded soil bunds combined with grass strips. These land units are characterized by slope class of

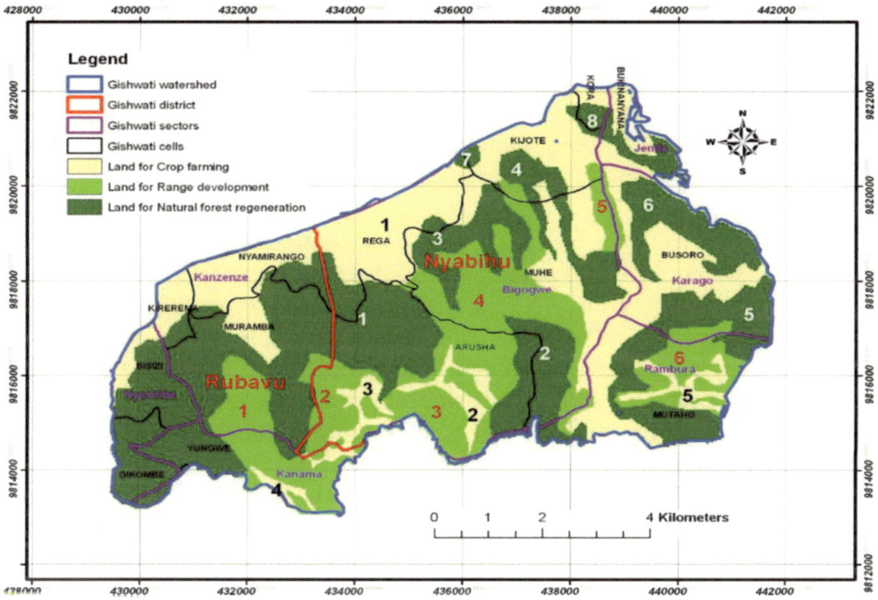

Figure 7.
Land management blocks grouping identified land units and boundaries of the different levels of administration. Source: GWLM project.

Soil depth	Land units (ha) by slope classes, soil depths and soil types					
	0–6%	6–16%	16–40%	40–60%	>60%	Total
Rock	0.36 (16)	1.53 (17)	12.42 (18)	21.06 (19)	3.51 (20)	38.88
0 - 50 cm	45.36 (7)	16.74 (8)	10.8 (11)	9.72 (12)	0.54 (15)	83.16
50 - 100 cm	256.77 (2)	144.45 (4)	278.46 (6)	143.37 (10)	34.83 (14)	857.88
100 - 200 cm	347.04 (1)	727.83 (3)	985.75 (5)	1316.43 (9)	240.48 (13)	5617.53
Grand Total	649.53	890.55	3287.43	1490.58	279.36	6597.45

Table 1.
Land units of different management and land uses. Source: GWLM project.

0–6% and soil depth of 0.5 to greater 1 m deep and are more productive for annual cropping with relatively less expensive land mangement measures. Land units 3 and 4 on slope range of 6–16% and soil depth of 0.5 to 1 m deep or more were treated with graded bench terraces integrated with agroforestry species. The embankments were protected by Kikuyu grasses. These 4 land units embraced crop farming but also some settlement places.

Land unit 5 was allocated to rangeland development using pasture grass (Kikuyu grass, *Phalaris aquatica*, etc), and forage legumes to feed the livestock. This unit was located on slope class of 16–40% and slope depth greated than 100 m but are underlain by rock surface to cause landslide problem when tree planting or continued cultivation is practiced.

Land units of 6–20 were allocated for natural forest regerations as they are strongly constrained either by absence of soil depth or excessive slope gradient (greater than 60%) and fragile soil. Land Unit 6 was constrained by the combined effect of the rolling topography (16–40%) and the shallow soil depth (50–100 cm). Land units 13–15 were in slope range exceeding 60% with more than 1 deep soil to cause landslide if no natural forest regeneration is applied. Land units 9 and 10 located at slope classe of 40–60% and majorly soil depth greater than 1 m would be prone to landslide. The assignment of land units 7 and 8 were linked to shallow depth (0–50%) while the land units 16–20 were allocated to this land use becaused exposed rock. Natural forest regeneration and restoration covered about 2970 ha in these land units. The implemented landscape restoration interventions were accompanied by drainage system of water ways, cut-off drains, agroforestry systems and live fences (rangeland).

Finally, three blocks were formed to group land units with similar land use. Land units 1–4 suited for crop farming while land unit 5 was assigned to rangeland development. Land units from 6–20 matched for natural forest regeneration. With use of GIS tools, concrete pillars were installed demarcating different land management blocks (Cropland, Pastureland and Forestland). However, land use category that occurs in less than 8 ha was not considererd to stand as a block by its own but it was annexed to adjacent land unit for ease of mangement and practicality of implementation point of view.

According to this harmonized block formation, the lands recommended to be put under natural forest regeneration covered about 45% whereas lands for range development and cultivation covered 23.3 and 31.6% respectively. This land use planning helped to not only guide the implementation of appropriate husbandry technologies but also for better allocation and management of resources. As discussed above, farmers participated in the identification at the extent they got informed about the specificity of interventions in their farms and cross-boundary conditions in the context of land consolidation (**Figure 8**).

3.3.2.2 Cropland blocks

Croplands were either subjected to graded terraces connected to cut-off-drains and water ways or minor agriculture intervention with graded soil bunds for about 2087 ha. Among others, coverage area of 1654 ha has been terraced and protected against erosion and floods as it is illustrated in the **Figure 9**. Interventions also included biological measures such as grasses, trees and herbaceous legumes.

Pyrethrum growing activities have successfully been established in Gishwati area under rotation system with Irish potatoes. It contributed to the increase of the national area for pyrethrum cash crop. To this effect, 102 hectares have been planted with pyrethrum in Gishwati which served as seedlings to the areas outside

Figure 8.
Land demarcation and installation of the benchmarks for land consolidation purpose. Source: GWLM project.

Gishwati. At side, nurseries were established at 30 ha for supplying good quality of pyrethrum seedlings.

3.3.2.3 Pasture/rangeland blocks

Degraded lands have been converted to prescribed pasture/rangeland blocks for an area of 1540 ha by planting kikuyu grass (**Figure 10**). This has been supplemented with silvo-pastoral activities.

3.3.2.4 Forestland blocks

Land allocated for natural forest regeneration within forestland blocks received both exotic and indigenous tree species at 2970 ha. Tree planting has been sustained with constant monitoring to protect against grazing and prematured harvesting (**Figure 11**).

3.3.2.5 Complementary engineering works for Gishwati watershed protection

Additional engineering works were constructed to deal with the flooding and poor drainage problems. They comprised the construction of Kinamba Bridge

Figure 9.
Landscape management using bench terraces. Source: GWLM project.

Figure 10.
Degraded land with and without rangeland development. Source: GWLM project.

Figure 11.
Degraded lands restored with tree planting activities for forest regeneration. Source: GWLM project.

Figure 12.
Construction of bridge and river drainage canal. Source: GWLM project.

along with strengthening roadside channels, retaining walls, filling and compaction of main road with gravel soil (**Figure 12**). In addition, drainage rehabilitation of Mizingo River was reinforced with stone masonry to protect flooding in the lowland.

4. Land husbandry interventions within an integrated watershed management approach

4.1 Description of the study areas

Land husbandry interventions that are suitable for hilly landscape were extensively introduced in the country since 2010 to control erosion and runoff. This strategic action has been initiated by the Land Husbandry Water Harvesting and Hillside irrigation (LWH)[1] project under the MINAGRI to

[1] Project funded by the Government of Rwanda and multi-donor organizations such as USAID, the World Bank, the GAFSP, and the Canadian International Development Agency.

boost the land productivity. The purpose was to introduce a wide range of innovations for improving agricultural practices, sustaining land management conditions and combating food insecurity by increasing rural community's livelihoods income.

As precedently discussed, the LWH project lies its focus on modernizing agricultural farming activities in hilly landscapes subjected to erosion, fertility depletion, and acidity problem. The Nyanza and Karongi sites have been selected for solving such problems in rural farming system. Nyanza 23 site is located at latitude of 2.365618°S and longitude of 29.692154°E while Karongi 12–13 sites are located at latitude of 2.0530°S and longitude of 29.468052°E, and latitude of 2.043841°S and longitude of 29.492853°E, respectively.

4.1.1 Nyanza 23 characterization

Nyanza 23 site is located in the Nyanza District of Southern Province. The site covers a good portion of Rwabicuma, Nyagisozi and Cyabakamyi and small part of Busasamana sectors of Nyanza District and Rwaniro sector of Huye District. It covers 5,659 ha as illustrated in the **Figure 13**. It comprises an irrigation dam which is supplied by Gisuma and Gasenyi tributaries of Kagondo stream and irrigates the downward part.

Climatic data from the Rwanda Meteorological Agency (RMA) in Nyanza 23 show the mean annual rainfall of 1,177 mm per year with the driest and wettest months of July and April, 7 and 190 mm respectively. Rainy seasons last from March to June and October to December, alternating with dry seasons. Although Nyanza district generally exhibits moist rainfall conditions but on-site rainfall data showed deficit of water reducing the expected optimal crop yield. Mean temperature is excellent for plant growth but the evapotranspiration values indicated the need for additional water supply (irrigation).

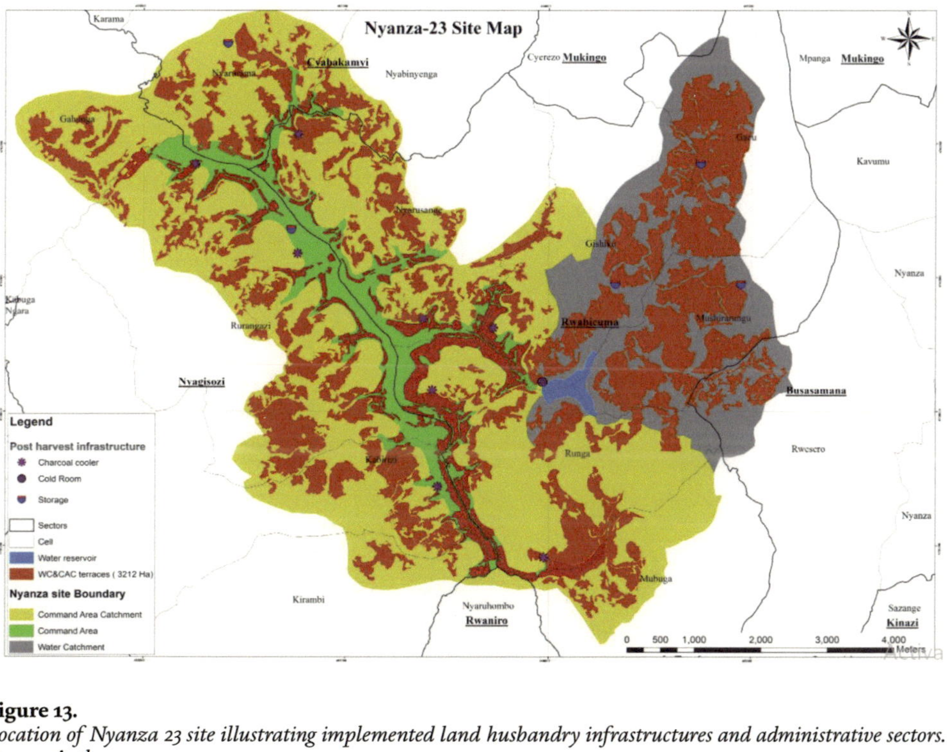

Figure 13.
Location of Nyanza 23 site illustrating implemented land husbandry infrastructures and administrative sectors. Source: Author.

In terms of topography, Nyanza 23 catchment illustrates five distinct slope categories using the methodology of the Digital Elevation Model (DEM–30 m resolution), accessed from the United States Geological Survey (USGS) database (http://earthexplorer.usgs.gov/). Slope gradient ranges from 0–6%, 6–16%, 16–40%, 40–60%, and > 60% that respectively covered the percentage area of 10.7, 30.0, 52.7, 6.0 and 0.61 of the catchment. The range between 16–40% dominates the study area and about 2/3 of this area has shallow soils. About soil characteristics, the catchment is dominated by *Leptosols, Lixisols, Alisols, Gleysols, Cambisols, and Ferralsols* [19, 20]. The catchment is generally dominated by coarse textured soils up to more than 60% of the total area whereas the remaining part is also moderately fine textured soil.

4.1.2 Karongi characterization

Karongi 12 and 13 sites are located in Rubengera, Rugabano and Mukura sectors of Karongi district. They respectively cover 651.3 and 226.2 hectares (**Figure 14**). The two sites fall in the moist mid-highland agro-climatic zone, which of great potential for agriculture. The altitude varies from 1940 to 2160 m in the catchment while slope gradient ranges from 4 to 71% across the catchment [19, 22]. The dominance of hilly topographic features in the area coupled with soil susceptibility accelerates erosion, thus land-husbandry in this watershed was crucially essential.

According to Rwanda Meteo Agency (RMA), the annual rainfall of the area is around 1300 mm, also expressing two wet seasons from September to December and March to June, respectively. Mean annual temperature is more or less than 18 °C. Although the area does not express the rainfall drought with 10% higher than annual potential evapo-transpiration, shortage of rainy seasons and problems of dry spells drastically affect crop growth. Thus, it requires additional supply of water through irrigation.

The soils are deep with soil depth greater than 50 cm covering more than 90%. Soil types are *Humic Acrisols and Cambisols* on the hillside while in the valley

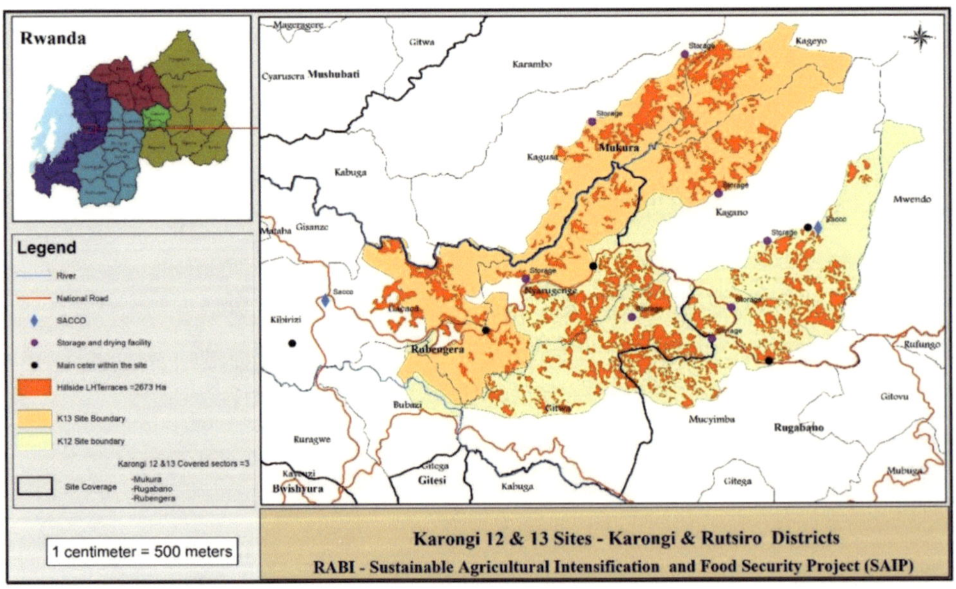

Figure 14.
Location of Karongi 12–13 sites illustrating implemented land husbandry infrastructures and administrative sectors. Source: Author.

bottom, *Umbric Gleysols* are present. Soils are dominantly medium textured classes (clay loam and sandy clay loam) with potential to hold more water and have relatively good agricultural potential [19, 20].

4.2 Approach

The development of the two catchments followed a participatory integrated watershed management involving farmer's community's contribution and landscape-based interventions. Socio-economic aspect considered the responsiveness of beneficiaries, local authority, gender aspect and expecting site-specific economic rate of return. On the other side, technical aspect lies on severity erosion towards environmental impact of the catchment protection, and potentiality for hillside irrigation on developed land husbandry works (terraces).

Therefore, the catchment was divided into the command area locates in the downward part of the constructed dam and the catchment area which is the hillside surrounding the dam at upstream part (**Figure 15**). Hillsides of both sites are protected against erosion risks with appropriate erosion control measures, especially bench terraces. Terraces in the hillside surrounding the downstream part (command area) are irrigated by water from the dam for increased more number of cultivation times compared to rainfed conditions. Terraced lands under irrigation will allow them to cultivate for three (3) agricultural seasons per year. Extensive community sensitization and participatory approaches ensured that farmers fully participated in their own transformation.

4.2.1 Implementation approach

The approach introduced comprehensive sustainable land husbandry technologies for soil erosion control and increasing soil fertility to boost the land productivity as well as develop water retention dams for hillside irrigation. It is considered as an active process of selecting and implementing systems of land use and management in such ways that there will be an increase in or at least no loss of land quality, soil health and land productivity. The implemented land husbandry interventions respected the participatory watershed-based approach using both erosion control

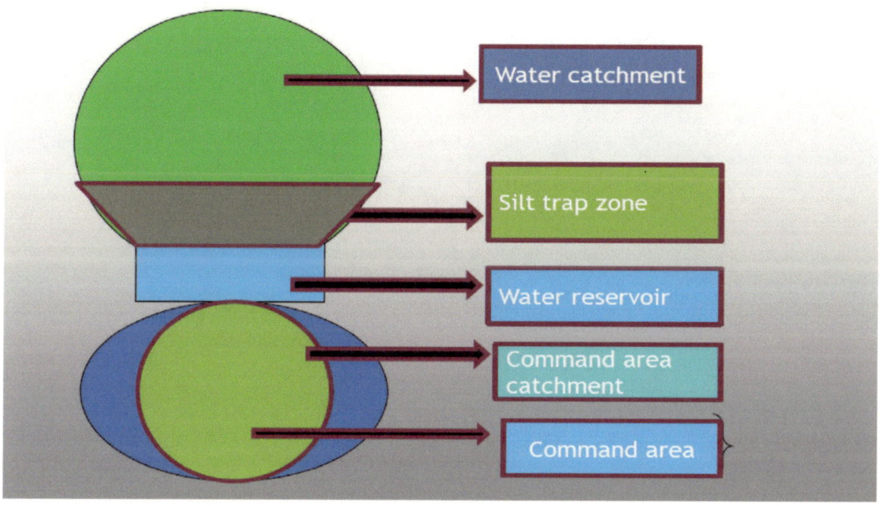

Figure 15.
Framework of landscape restoration under the LWH project. Source: LWH project.

measures and effective use of soil amendments (lime and compost). The sequencing of implementing activities were as follow: mobilization of staff and local authorities, mobilization of labor (mainly beneficiaries), training of labor on land husbandry technologies, and implementing land husbandry works.

Land husbandry technologies included grass strips, trash lines, earth/stone bunds, bench terraces, protected cut-off drains, water ways, gully plugs, embankment shaping, narrow-cut terraces, pitting, and conservation ridges/ditches as illustrated in the **Figure 16**. These are supplemented by the use of composting, mulching, liming and green manuring applications [14]. These land husbandry technologies have started on the upper side of the hill where the slope is under 6% where the first cut-off drain is located. Below this cut-off drain, other comprehensive land husbandry technologies are applied depending on land use, slope category and agro climatic zones.

Distribution of agro-climatic zones across the country influenced the types and forms of measures (**Table 2**). The wet agroclimatic zones have high rainfall amount of 1400 mm per, that increases its intensity as altitude increases and significantly causes flood, siltation and landslide. Therefore, the choice of land husbandry technologies follows the capacity to obstruct erosive force by an an integrated physical and biological measures, discourage water movement from attaining maximum velocity, improve conditions for surface drainage where infiltration causes landslide, and finally drain water from drained fields to safe storage such as valley dams, cascade ponds, rivers and large drainage canals. Graded bench terraces connected to cut-off-drains and waterways are developed towards reservoirs or river.

The moist agroclimatic zones with annual rainfall amount between 900 and 1400 mm per year require tailored land husbandry measures as leveled bench terraces and contour bunds interspaced by cut-off drains that convey excess of water to water-ways during rainy seasons and finally into a reservoir or water body. The agroclimatic dry zones (<900 mm per year) are characterized by low rainfall that needs land husbandry measures for retaining moisture. The leveled structures (terraces) with tie-ridges are recommended to help supplementary water supply.

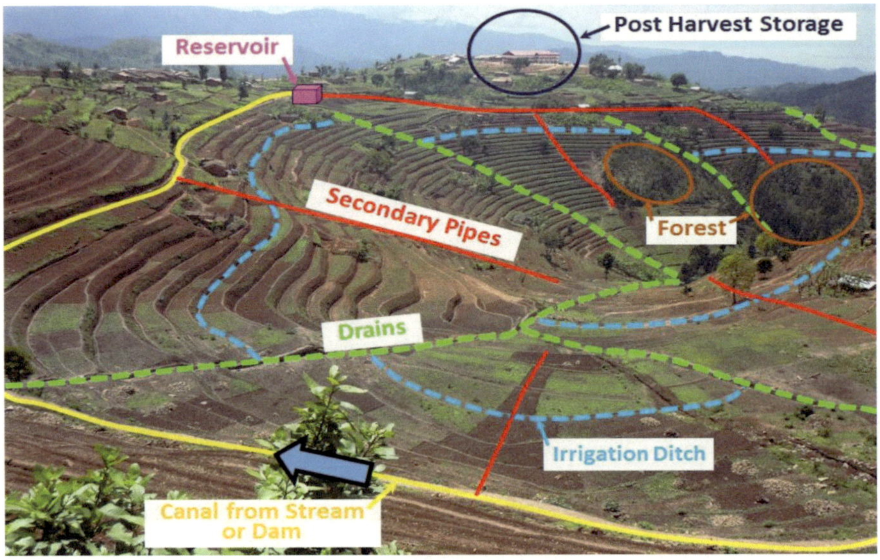

Figure 16.
Demonstration of land husbandry implementation. Source: LWH project.

Slope categories (%)	Types of bench terraces	Soil depth in (cm)	Vertical interval (m)	Spacing (m)
16–40	Leveled Bench Terraces	75	1.5	9.4–3.7
20	Idem	75	1.5	7.5
39	Idem	75	1.5	3.8
40–60	Narrow cut- Bench Terraces	100	2	
45	Idem	100	2	4.44
59	Idem	100	2	3.4
Greater than 60	No Bench Terraces are implemented	—	—	—

Source: adapted from Bekele-Tesemma [14].

Table 2.
Specification of some technical guidance for construction of bench terraces.

4.2.2 Technical specifications of terrace establishment

Establishment of well-established terraces is meant to follow technical specifications linked to slope gradient, soil depth, and soil types [14]. They provide technical guidance about how terraces are technically constructed, maintained, and cultivated. The technical recommendations of bench terraces are based on an assumption of a soil depth of between 75 cm and 1 m and Vertical Interval (VI) of between 1.5–2 m and also the calculation counted the Vertical Interval (VI) for the space needed between two succeeding bench terraces. Computation may vary depending on whether bench terraces are being constructed using machines or hand-made (Mesfin [23]).

$$VI = \frac{S * WB}{100 - (S * U)}$$

Where *VI*: Vertical interval in m; *S*: Slope in percent (%); *WB*: Width of bench (flat strip) in m; *U*: Slope of riser (using value 1 for machine-built terraces, 0.75 for hand-made earth risers and 0.5 for rock risers).

In the **Table 2** shows the comprehensive guidelines for soil erosion control measures based on slope, soil type, depth and agro-climate [14].

4.3 Results of successful land husbandry interventions

4.3.1 Technical achievements

Successful results covered more than 19,500 ha with comprehensive land husbandry technologies across the country out of which over 3,400 ha were located on marginal lands. The lands were made productive after land husbandry works. For this particular study cases, bench terraces were established at 2673 (gross area of 4284 ha) and 3212 ha (gross area of 4800 ha) of lands for Karongi 12–13 and Nyanza 23, respectively (**Figures 13** and **14**). Technologies effectively reduced erosion for about 98% of the total soil losses. Other land uses such as forest, settlements, water reservoirs and papyrus were also rehabilitated on about 901 ha.

Besides the use of land husbandry technologies, the sites were restored in terms of soil fertility replenishment through the use of lime (5 t ha^{-1}), compost/manure

(10 t ha^{-1}) and mineral fertilizer (DAP and Urea) inputs accompanied with irrigation in the command area for optimally increasing productivity. Farmers are growing food crops like beans, maize, cassava, sweet potato, sorghum, banana, vegetables (chili, tomato, eggplant, onion, sweet pepper) and various fruit species (water-melon, tree tomato, avocado, macadamia, etc). However, the productivity did not reach the expected optimal yield.

4.3.2 Social build-up of farmers exploiting developed lands

Participatory consideration was also a key in the successful of sustainable landscape interventions implemented in the degraded lands. At first, farmers have generated more income from labor works of establishing comprehensive land husbandry measures. Through this process, farmers in which 47% were female, earned income which helped to finance their livelihoods through financing facilities. In addition, Communities were grouped into self-help groups (10 persons) building into zones which lead to cooperative formation. Cooperatives were created and strengthened through various trainings to sustainable manage and valorize the established land husbandry works.

As farmers grow several crops, it was worth to built post-harvest handling facilities to reduce postharvest losses while strengthening crop value chains and marketing systems such as storages facilities, drying shelters, collection centers (for banana), horticulture collection centers including charcoal coolers, temporary drying facilities constructed during harvesting seasons, and other necessary equipments. Briefly, activities have not only included the technical aspects but also community sensitization to ensure that people fully participate in their own transformation. This wide range of capacity building initiatives were also supported agriculture and extension services (Districts ...).

5. Lesson learnt and discussion

5.1 Success stories for participatory landscape management in Gishwati area

Gishwati area was restored in a participartory landscape approach within planning and implementation processes at 6,600 ha. It comprises activities of land husbandry on agriculture land, reforestation, and rangeland rehabilitation. The approach also considered the relocation of people from high risk zones to other places and building capacities of farmers through farm-livestock cooperatives. Thus, this approach has successfully facilitated to establish a comprehensive landscape management to effectively address the frequent landslides and flooding and sustainably exploit the land to the profit of local farmers in the livelihoods and the country's economy in general.

The evidences demonstrated how land husbandry interventions within participatory landscape approach especially terraces are very efficient not only in technical aspects of controlling soil erosion and boosting productivity but also improving people's livelihoods. According to Rutebuka et al. [13, 24] in Rwanda, bench and progressive terraces effectively control erosion up to 90% of soil and nutrient losses, once they are well established, managed and regularly maintained by landowners (farmers). The study in Ethiopia highlands substantiated the impact of terracings which reduced loss of soil from 97 to 38 t ha^{-1} yr^{-1} during 1984 and 1988 in Minchet catchment [25].

The Government for the sake of promoting agriculture and natural resource management has effectively addressed the challenges linked to bio-physical (land size, erosion, climate, and acidity), structural, and institutional contexts. The success stories resulted from planned land use, served in solving land related issues. High value indigenous tree species have been re-introduced in the area for the purposes of rehabilitating the ecosystem of Gishwati and developed lands have been effectively redistributed to beneficiaries. To sustain the established land husbandry works required a process of building capacity of people for increasing ownership and commitment of land beneficiaries. It comes into practice through HIMO approach through community sensitization, exploring social relations, monitoring of implemented works, and protecting grazing lands in restricted high risk zones. HIMO approach also created employment to more beneficiaries.

5.2 Success and challenges for establishing and managing terraced lands under LWH development

LWH project development also demonstrated how land husbandry technologies especially bench terraces are technically efficient in soil erosion control wherever they were well established, managed and maintened. Comparing before and after establishment of land husbandry technologies, the rate of soil erosion has been reduced from 50–100 t ha^{-1} yr^{-1} in 2011 to less than 50 t ha^{-1} yr^{-1} in 2014 as reported in the LWH project report in one of the project site of Rwamagana district (**Figure 17**). This is also confirmed by Rutebuka [8] in the study site of Rwamagana district developed by LWH project that bench terraces reduced soil loss from 23.5 to 1.7 t ha^{-1} yr^{-1} in the catchment landscape with slope gradient varying between 0–60%. In Ethiopia highland, terracing techniques controlled soil erosion by 39.1% in the period of four years (1984–1988) [25].

The erosion control is not an end itself, but cropland has to provide expected ecosystem benefits, of which the increase in crop productivity is a paramount. Development of land by terraces increased production of crops compared to what was before. Implementation of integrated land husbandry technogies changed the livelihood conditions of the poorest areas through modernizing agricultural

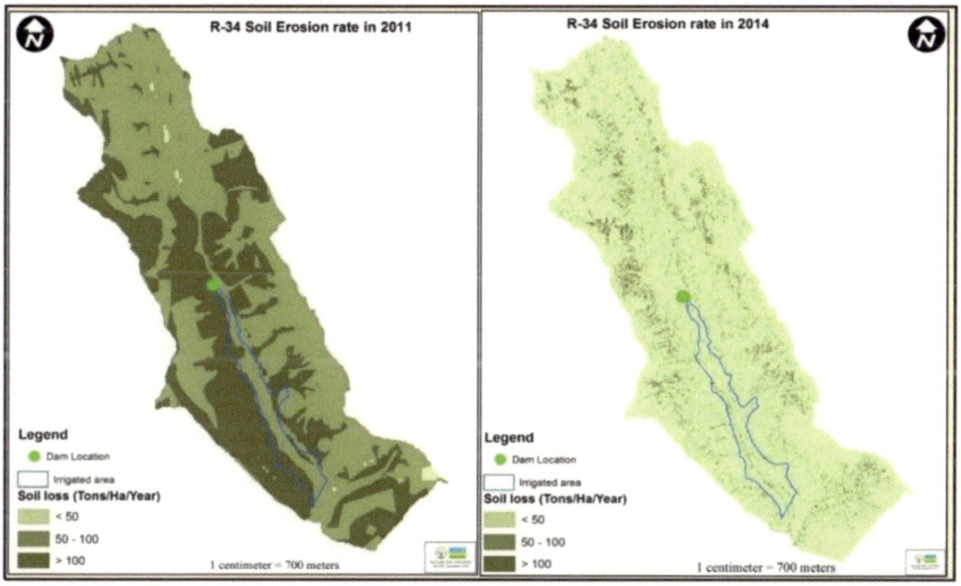

Figure 17.
Change in soil loss before and after development of land husbandry technologies at Rwamagana 34 site under LWH project. Source: LWH project.

techniques and increasing income levels. Hundreds of thousands of poor rural farmers in the project intervention areas have been supported to break out of poverty and obtain food security. HIMO approach within an integrated participatory watershed management contributed to the creation of jobs and reinforcement of farmer's capacity in during implementation of land husbandry technologies. HIMO provides benefits of promoting employment, organizing farmers into community groups, using local resources such as supplying of organic materials, increasing knowledge and skills of local farmers and offering people access to income and financial schemes (Banks, saving schemes). However, farmers were unable to reach the optimal production potential, as a result many rural farmers barely produced enough to feed their families.

Concerning the study cases of Nyanza 23 and Karongi-12 & 13, it was expected to continue increasing agricultural productivity from this comprehensive land husbandry technologies. Unfortunately, some developed terraces in the case studies have been affected by low productivity of crops, resulting from both under-exploitation and abandonment problems of terraced lands [26]. Productivity problems could originate from the way terraces have been constructed on very acidic and inherently unfertile soils with inadequate supply of organic manure, fertilizers, lime and other land related problems [24, 27, 28].

5.3 Lesson learnt from Rwanda experiences in land husbandry

The same issue was also observed on terrace construction through collective actions such as VUP (Vision 2020 Umurenge Program) or other service providers from the District initiatives. In this case, low productivity is not only related to low productivity but also the establishment approach. Concerns are when the service providers might be driven by the completion rate of the contract signed by compromising technical guidelines like saving the top and nutrient soils during terrace construction, slope and soil types as well as not adopting a participatory integrated watershed management approach.

Another aspect hindering the success of terrace development relies on social-economic context. Farmers might be reluctant in adopting land husbandry technologies like terraces if they are not getting expected optimal yield in the first years because it requires at least four years for restoring soil fertility. The low understanding may result in low efficiency of terrace exploitation [29]. These factors relate on economical and institutional aspects along the implementation of bench terraces that are likely to constrain future use and maintenance of these structures [30]. Higher costs of investment and maintenance compared to the farmer's capacity hindered farmers to exploit these terraces.

Recent study identified problems affecting the poor performance of developed lands due to both technical and socio-economic aspects [31]. The findings proposed possible and best options to ensure that the lands are being optimally utilized for improving crop productivity. It includes improvement of soil fertility with supply of lime and organic amendments, agronomic practices and intensifying agroforestry systems for under-exploited or abandoned terraced lands. At least 2.5 t ha^{-1} of lime should be applied for soil acidity with pH less than 5.5 while 10 t ha^{-1} of organic manure of good quality has to be applied at every cropping season. Apart being well established, socio-economic challenges have to be well addressed by organizing or strengthening cooperatives of farmers and provide financial and technical supports that could help to alleviate identified financial barriers.

All these factors may result in unstable terraces that could accelerate the accumulated runoff volumes, from the destruction of risers and more eroded materials [24]. At some extent, these abandoned terraces can cause landsliding, mass movements and gullies [32–34]. Thus, it is required to enforce the updated technical guidelines and standards for well-established terraces within an integrated and participatory landscape approach.

6. Conclusion

This chapter described different erosion control approaches that have been adopted in Rwanda, focusing on two selected case studies such as Gishwati area and LWH project sites (Karongi and Nyanza). It pinpoints the success stories in land husbandry interventions that can be scaled up to other regions with similar landscape properties. Challenges observed can also serve as lessons learnt in future interventions within or outside of Rwanda.

Participatory landscape approach promoted in Gishwati area was a success story in protecting degraded lands and generating ecosystem benefits. The more integrated natural resources management, and participatory planning helped for addressing the frequent landslides and flooding while sustainably exploit the land to the profit of local farmers in the livelihoods and the country's economy in general. This approach comprises development of agriculture land, reforestation, and rangeland rehabilitation, relocation of people from high risk zones and building capacities of farmers through farm-livestock farmers's organization.

On the other hand, the LWH projects provided strong evidences how land husbandry technologies (terraces) efficiently reduced erosion risks and improved farmers' livelihoods through crop productivity increase. However, it also highlighted the challenges observed in the adoption of integrated watershed management which did not tackle some technical and socio-economic aspects. Technical problems could result from inappropriate establishment of terraces without incorporating recommended technical guidelines related to soil types, depth and slope. These resulted into terrace destruction leading to mass movements, gullies and siltation in the valleys. Socio-economic challenges importantly cause farmers for abandoning or under-exploiting terraced lands. Terraces on very acidic and inherently unfertile soils require an intensive supply of organic and lime amendments together with use improved agronomic practices and agroforestry systems.

Finally, this chapter recommends the land husbandry policy strategies to successfully adopt the participatory landscape management for optimizing land's productivity in a sustainable manner. Ths involves the participation of farmers' communities from planning up to the implementation processes as well as valorization of terraced lands. HIMO approach is also suggested in the development of rural communities. Farmers should be grouped in rural communities (cooperatives) to increase their financial and technical skills.

Acknowlegment

I would like to say many thanks for the support from the former staff under GWLM and LWH projects who provided to me project materials and at some extent physical/virtual contacts we have had together for improving and finalizing this chapter.

Author details

Jules Rutebuka
Rwanda Agriculture and Animal Resources Development Board (RAB), Kigali City, Rwanda

*Address all correspondence to: rujub2@gmail.com

References

[1] NISR, 2018. Seasonal Agricultural Survey 2018 annual report. National Institute of Statistics of Rwanda (NISR). Kigali, Rwanda. http://statistics.gov.rw/datasource/seasonal-agricultural-survey-2018.

[2] NISR, 2019. Rwanda Statistical YearBook 2019. National Institute of Statistics of Rwanda (NISR). Kigali, Rwanda. http://statistics.gov.rw/publication/statistical-yearbook-2019.

[3] Karamage, F., Zhang, C., Ndayisaba, F., Shao, H., Kayiranga, A., Fang, X., Nahayo, L., Nyesheja, E.M., Tian, G., 2016. Extent of cropland and related soil erosion risk in Rwanda. Sustain. 8, 1-19. https://doi.org/10.3390/su8070609.

[4] MINAGRI, 2013. Strategic Plan for the Transformation of Agriculture in Rwanda Phase III (2013-2017). Kigali, Rwanda.

[5] RAB, 2012. Soil erosion control baseline in Rwanda. : Ndabamenye, T, MUsana S.B., Ngoga T.G., Kagabo M.D., Dusengemungu L., NABAHUNGU L.N., and Mbonigaba J.J.M.

[6] Kagabo, D.M., Stroosnijder, L., Visser, S.M., Moore, D., 2013. Soil erosion, soil fertility and crop yield on slow-forming terraces in the highlands of Buberuka, Rwanda. Soil Tillage Res. 128, 23-29. https://doi.org/10.1016/j.still.2012.11.002.

[7] Rushemuka, P.N., Bizoza, R., Mowo, J., Bock, L., 2014. Farmer's soil knowledge for effective participatory integrated watershed management in Rwanda: Toward soil specific fertility management and farmers' judgemental fertilizer use. Agric. Ecosyst. Environ. 183, 145-159. https://doi.org/10.1016/j.agee.2013.10.020

[8] Rutebuka, J., Kagabo, D.M., Verdoodt, A., 2019. Farmers' diagnosis of current soil erosion status and control within two contrasting agro-ecological zones of Rwanda. Agric. Ecosyst. Environ. 278, 81-95. https://doi.org/10.1016/j.agee.2019.03.016.

[9] Mutwewingabo, B., Rutunga, V., 1987. Projet d'intensification agricole, PIA Gikongoro. Kigali, Rwanda.

[10] Roose, E., Ndayizigiye, F., 1997. Agroforestry, water and soil fertility management to fight erosion in tropical mountains of Rwanda. Soil Technol. 11, 109-119. https://doi.org/10.1016/S0933-3630(96)00119-5.

[11] Ngarukiyimana, J.P., Fu, Y., Yang, Y., Ogwang, B.A., Ongoma, V., Ntwali, D., 2018. Dominant atmospheric circulation patterns associated with abnormal rainfall events over Rwanda, East Africa. Int. J. Climatol. 38, 187-202. https://doi.org/10.1002/joc.5169.

[12] Ntwali, D., Ogwang, B.A., Ongoma, V., 2016. The impacts of topography on spatial and temporal rainfall distribution over Rwanda based on WRF model. Atmos. Clim. Sci. 06, 145-157. https://www.scirp.org/journal/PaperInformation.asp...

[13] Rutebuka, J., De Taeye, S., Kagabo, D., Verdoodt, A., 2020a. Calibration and validation of rainfall erosivity estimators for application in Rwanda. Catena 190, 104538. https://doi.org/10.1016/j.catena.2020.104538.

[14] Bekele-Tesemma, A., 2011. Illustrated supervision checklist for assessment of the quality of comprehensive land-husbandry works at LWH project Sites in Rwanda. Kigali.

[15] Ngarambe, V., 2004. Plan stratégique de transformation de l'agriculture au Rwanda: Gestion et utilization de l'eau et des sols, Document préparé du Groupe d'Expertise, de

Conseil et d'Appui au Développement (GECAD). Kigali, Rwanda.

[16] Dressler, J., 1983. La structure agraire de la région du projet. Résultats d'enquête socio-économique. Études et Expériences n°2.: PAP Nyabisindu. Rwanda.

[17] Steiner, K.G., 1998. Using farmers' knowledge of soils in making research results more relevant to field practice: Experiences from Rwanda. Agric. Ecosyst. Environ. 69, 191-200. https://doi.org/10.1016/S0167-8809(98)00107-8.

[18] FAO, UNESCO, ISRIC, 1990. Revised legend of the soil map of the World. World soil resources report. Rome, Italy.

[19] Imerzoukene, S., Van Ranst, E., 2001. Une banque de données pédologiques et son SIG pour une nouvelle politique agricole au Rwanda. Bull. Séanc. Acad. R. Sci. O.-m. 47, 299-329.

[20] Verdoodt, A; Van Ranst, E., 2006. The soil information system of Rwanda: a useful tool to identify guidelines towards sustainable land management. Afrika Focus, Vol. 19, Nr. 1-2, 2006, pp. 69-92. https://doi.org/10.21825/af.v19i1.5415.

[21] Tantoh, H.B., Simatele, D.M., Ebhuoma, E., Donkor, K. and McKay, T.J., 2019. Towards a pro-community-based water resource management system in Northwest Cameroon: practical evidence and lessons of best practices. GeoJournal, pp.1-19. https://doi.org/10.1007/s10708-019-10085-3.

[22] Delepierre, 1974. Les régions agricoles du Rwanda. Note technique No. 13. Rubona, Rwanda.

[23] Mesfin Akilu, 2016. A field guideline on the bench terrace design and construction. Ministry of agriculture and Natural resources. Addis Ababa, Ethiopia. Addis, Ethiopia.

[24] Rutebuka, J., Munyeshuli Uwimanzi, A., Nkundwakazi, O., Mbarushimana Kagabo, D., Mbonigaba, J.J.M., Vermeir, P., Verdoodt, A., 2020b. Effectiveness of terracing techniques for controlling soil erosion by water in Rwanda. J. Environ. Manage. 277, 111369.

[25] Subhatu, A., Lemann, T., Hurni, K., Portner, B., Kassawmar, T., Zeleke, G., Hurni, H., 2017. Deposition of eroded soil on terraced croplands in Minchet catchment, Ethiopian Highlands. Int. Soil Water Conserv. Res. 5, 212-220. https://doi.org/10.1016/j.iswcr.2017.05.008.

[26] MINAGRI, 2015. Pilot exercise of profiling of developed terraces. Final report under Ministry of Agriculture and animal resources. Kigali, Rwanda.

[27] Rushemuka, N., 2011. Imperative combination of lime, organic matter and fertilizers to foster the productivity of the acid and depleted soils of Southern Rwanda. In: Woomer P.L., ed. Integrated soil fertility management in Africa: from microbes to markets. Kenya.

[28] Rutebuka J., 2019. Integrated multiscale assessment of soil erosion and its control in Rwanda. PhD thesis. [WWW Document]. Ghent Univ. Fac. Bisocience Eng. URL https://biblio.ugent.be/publication/8620562 (accessed 10.29.19).

[29] Kessler, C.A., 2006. Decisive key-factors influencing farm households' soil and water conservation investments. Appl. Geogr. 26, 40-60. https://doi.org/10.1016/j.apgeog.2005.07.005.

[30] Kees Burger; Fred Zaal, 2009. Sustainable Land Management in the Tropics: Explaining the Miracle (International Land Management):

Zaal, Fred, Burger, Kees: 9780754644552: Amazon.com: Books.

[31] Ruganzu, V., Rutebuka, J., Habamenshi, D., Munyeshyaka, J.D., Muhutu, C., Andre, C., Mwungura, M., 2020. Soil-crop suitability study of Nyanza 23 Site. Rwanda Agriculture and Animal Resources Development Board (RAB).

[32] Arnáez, J., Lana-Renault, N., Lasanta, T., Ruiz-Flaño, P., Castroviejo, J., 2015. Effects of farming terraces on hydrological and geomorphological processes. A review. Catena 128, 122-134. https://doi.org/10.1016/j.catena.2015.01.021.

[33] Inamdar, S.P., Dillaha, T.A., 2000. Relationships between drainage area, slope length, and slope gradient for riparian slopes in Virginia. Trans. ASAE 43, 861-866. Doi: 10.13031/2013.2981.

[34] Yongmei, D., Xihuan, S., Xianghong, G., Shijun, N., Juanjuan, M., 2011. Analysis of slope length on water soil erosion. In: 2011 International Conference on Consumer Electronics, Communications and Networks, CECNet 2011 - Proceedings. pp. 2943-2946. DOI: 10.1109/CECNET.2011.5769391.

Chapter 7

Biochar: A Sustainable Approach for Improving Soil Health and Environment

Shreya Das, Samanyita Mohanty, Gayatri Sahu, Mausami Rana and Kiran Pilli

Abstract

Current agriculture faces multiple challenges due to boom in food demand and environmental concerns. Biochar is increasingly being recognized by scientists and policy makers for its potential role in carbon sequestration, reducing greenhouse gas emissions, renewable energy, waste mitigation and as a soil amendment. The purpose of this review is to provide a balanced perspective on the agronomic and environmental impacts of biochar amendment to soil. Application of biochar to soil can play a significant role in the alteration of nutrients dynamics, soil contaminants as well as microbial functions. Therefore, strategic biochar application to soil may provide agronomic, environmental and economic benefits. Recent findings also supported that in order to enhance crop yield, improve soil quality and soil health, biochar has proven significant role as fertilizer and soil conditioner respectively.

Keywords: biochar, carbon sequestration, soil conditioner, waste mitigation, crop yield

1. Introduction

Agriculture plays an important role in shaping the global economy. Now-a-days, food security is a major issue. Despite remarkable refinement of agricultural practices after World War II, the global food supply is yet incapable to fulfill the actual demands. Further, emerging issues of soil pollution, climate change and-Desertification still remains to be iron out for the agriculture sector [1]. The global food demand is anticipated to increase by 70% till 2050 with the burgeoning population [2] and meeting up this demand without compromising soil health and agroecosystem has turned into a big challenge in the agriculture sector. To meet the pressing demand for food; indiscriminate use of fertilizers, plant growth regulators, pesticides etc. has become a general practice. Their excessive use is a serious concern because of their adverse impact on the environment and the entire food chain.

Depletion in soil organic matter and soil nutrients, decline in agricultural productivity due to excess use of chemical fertilizers and changes in climate due to anthropogenic activities are posing great threats to the sustainability of agricultural production in the tropical regions. So it is becoming important to use organic fertilizer along with inorganic fertilizer for improving sustainability and maintaining soil health. Along with organic manures and composts, the use of biochar is quite a novel approach having potential benefits to both environment and agriculture as the former is a source of calcitrant carbon and the later contains recalcitrant form of carbon. Application of biochar to soil as a technique to improve the quality of soil has emerged in recent years. A common characteristic of biochar is that it comprises mainly stable aromatic organic carbon that cannot readily be returned to the atmosphere [3, 4]. The decomposition rate of biochar is 0.03% per year. Once it is applied, it is able to help in water and nutrient retention for next 5–8 years. Furthermore, biochar can reduce the risk of environmental pollutants (organic and inorganic) from soils by forming complexes or through sorption of organic compounds like herbicides [5].

2. Production of biochar

2.1 Feedstocks/raw materials

Various organic materials are suitable as feedstock for the production of biochar. Biochar can be produced with raw materials such as grass, cow manure, wood chips, rice husk, wheat straw, cassava rhizome, and other agricultural crop residues [6]. Agricultural wastes (bark, straw, husks, seeds, peels, bagasse, sawdust, nutshells, wood shavings, animal beds, corn cobs and corn stalks, etc.), industrial wastes (bagasse, distillers' grain, etc.), agroforestry (*Gliricidia* twig, *Eucalyptus* bark, *Pongamia* shell, *Eucalyptus* twig and *Leucaena* twig) and urban/municipal wastes [7, 8] have been extensively used, thus also achieving waste management through its production and utility. Hard wood biomass containing 10% moisture content is best for biochar production. After collecting hard woods, removal of barks can help to avoid lignin effects. Cellulose, hemicellulose, and lignin polymers are the principal components of biomass used for the biochar production.. Among these, cellulose has been found to be the prime component of most plant-derived biomasses, but lignin is also important in woody biomass.

2.2 Production process

Thermochemical conversion technologies are more popular than biochemical conversion technologies in case of biochar production as the rate of hydrogen production and yield are quite lower in the later. The former can further be divided into combustion, pyrolysis and gasification. Different thermochemical processes involved in biochar production are shown in **Figure 1**. Biochar which is obtained by slow pyrolysis from biomass waste (agricultural, municipal, animal, or industrial sources), is highly porous, fine-grained, carbon dominant product rich in paramagnetic centres having both organic and inorganic nature, with large surface area possessing oxygen functional groups and aromatic surfaces [9] with the primary goal of soil improvement. The pyrolysis temperatures generally employed ranges from 300 to 1000°C. Different types of pyrolysis along with their operating conditions are summarized in **Table 1**. In the absence of oxygen, Pyrolysis rapidly heats biomass, driving off carbon monoxide and hydrogen and turning the residue into biochar, a carbon rich solid. In this process, a mixture of volatile gases is released which can be captured and condensed into an energy-dense liquid called bio-oil. Further it can be refined into diesel and other hydrocarbon products. Recently, it has been reported that biochar obtained from the carbonization of organic wastes can be a substitute that not only influences the sequestration of soil carbon but also modifies its physicochemical and biological properties [15].

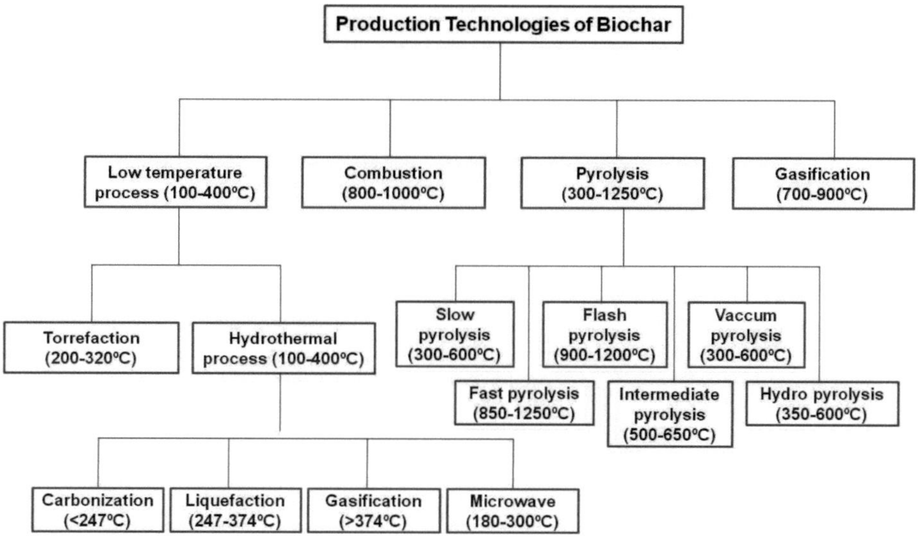

Figure 1.
Different thermochemical processes for biochar production.

Type of pyrolysis	Temp. (°C)	Heating rate (°C/s)	Pressure (MPa)	Residence time (s)	Particle size (mm)	Biochar yield (%)	References
Slow pyrolysis	300–600	0.1–1	0.1	300–550	5–50	20–40	Li et al. [10]
Fast pyrolysis	850–1250	10–200	0.1	0.5–10	<1	10–15	Li et al. [10]
Flash pyrolysis	900–1200	>1000	0.1	<1	<0.5	10–15	Li et al. [10]
Intermediate pyrolysis	500–650	1–10	0.1	10–20	1–5	15–25	Zhang et al. [11, 12]
Vaccum pyrolysis	300–600	0.1–1	0.01–0.02	0.001–1	—	25–35	Britt et al. [13]
Hydro pyrolysis	350–600	10–300	5–20	>15	—	—	Liu et al. [14]

Table 1.
Types of pyrolysis and their operating conditions for biochar production.

2.3 Factors associated with biochar quality

Specifically, the quality of biochar depends on several factors, such as the type of soil, metal, and the raw material used for carbonization, the pyrolysis conditions, and the amount of biochar applied to the soil (**Figure 2**).

The tendency of the surface functional groups to attract positive charges enhances the cation exchange capacity, which is an important property of biochar for remediation of metal-contaminated soils. The advantages of biochar with various physiochemical properties are shown in **Figure 3** [16].

The physical properties of biochar play significant role to its function as a tool for managing the environment. Research has been shown that biochar, when used as a soil amendment, improves soil quality and boosts soil fertility by increasing the

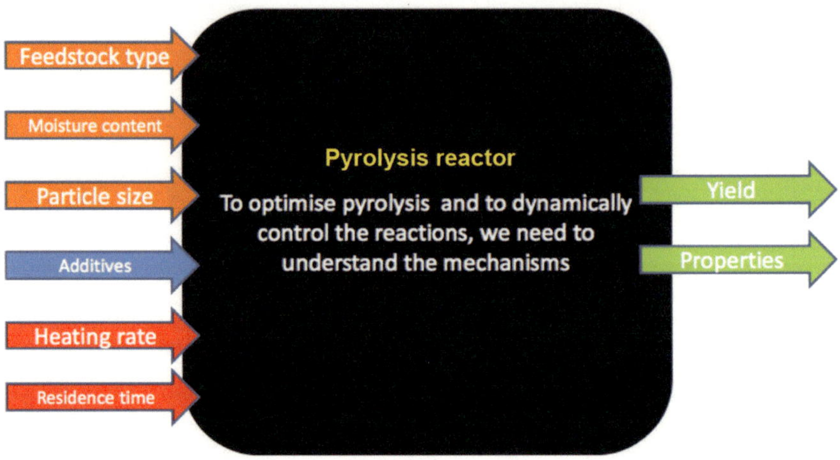

Figure 2.
Schematic diagram of factors affecting the quality of biochar.

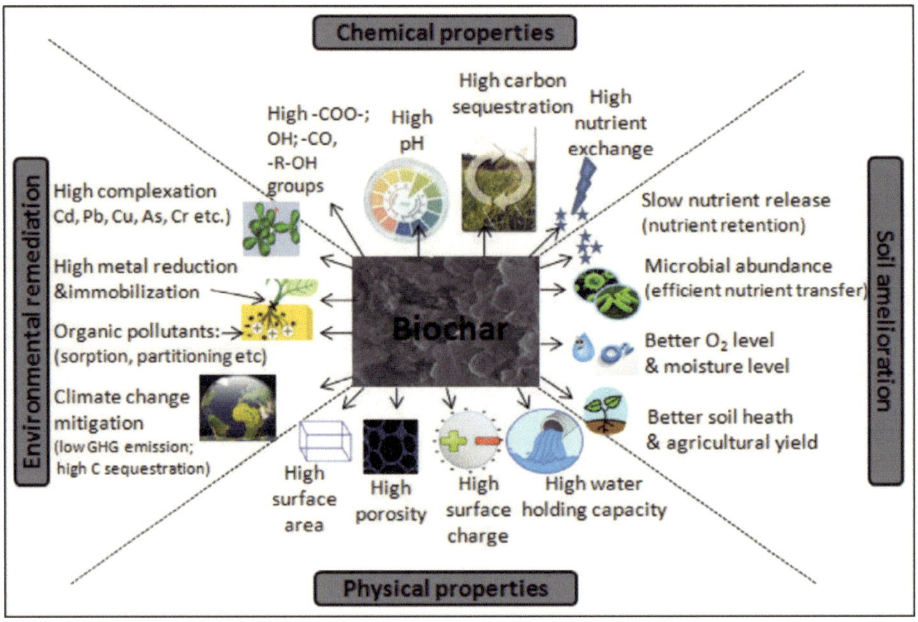

Figure 3.
Physicochemical properties of biochar.

moisture retaining capacity, soil pH, cation exchange capacity, attracting more beneficial fungi and other microbes, and preserving the nutrients in the soil. Biochar increases soil aeration and cation-exchange capacity, reduces soil hardening and soil density and changes the soil structure and consistency by changing the physical and chemical properties. In drought prone areas, the effects of drought on crop productivity can be reduced by addition of biochar due to its moisture-retention capacity. It has also been reported that it eliminates soil constraints that limit the growth of plants, and neutralizes acidic soil because of its basic nature [17].

As far as its chemical properties are concerned, biochar reduces soil acidity by increasing the pH (also called the liming effect) and helps the soil to retain nutrients and fertilizers. The application of biochar improves soil fertility through two mechanisms: adding nutrients to the soil (such as K, to a limited extent P, and many micronutrients) or retaining nutrients from other sources, including nutrients from the soil itself. However, the main advantage is to retain nutrients from other sources.

3. Effect of biochar on agricultural productivity, soil health and environment

Food security, climate change, declining soil fertility and profitability are the burning issues under the present scenario. Soil carbon is important for food security, ecosystem functioning and environmental health, especially in light of global climate change. Owing to its biological origin and physico-chemical properties, biochar (pyrolyzed crop residue) has the potential of carbon sequestration. Its higher stability against decay and capability to retain nutrients ensure the nutrient availability according to the crops need for a longer period.

3.1 Agricultural implications of biochar

Biochar has a diversed application ranging from use in agriculture and animal husbandry, flue gas cleaning, heat and power production, metallurgical applications, building material, to medical use. It has gained increasing popularity in the last years as a replacement for fossil carbon carriers in several of these applications in an attempt to reduce greenhouse gas emissions.

3.1.1 Biochar as a soil amendment

The burning issues as food security, climate change, declining soil fertility and profitability act as incentives behind the introduction of latest technologies of new farming systems. To reduce the risk of pollutant transfer to waters or receptor organisms in proximity, the amendment of soils for their remediation has proven a significant role.. In this context, the organic material such as biochar may serve as a popular choice owing to its biological source and direct application to soils with little pre-treatment. The two things which make biochar amendment superior to other organic materials are high stability against decay so that it can last for longer times in soil providing long-term benefits and high capability to retain the nutrients. Biochar amendment also play a significant role in improving soil quality by increasing moisture-holding capacity, soil pH, cation-exchange capacity and microbial flora [18]. The addition of biochar to the soil has shown the increase in availability of basic cations as well as in concentrations of phosphorus and total nitrogen [19, 20]. Another valuable property of biochar is suppression of emissions of greenhouse gases in soil. Due to the presence of calcium compounds, as well as improved physico-chemical and biological properties of soil, application of biochars and biochar amended composts is advocated to control the diseases caused by fungi and bacteria in soil. Bio-char can also adsorb pesticides, nutrients, and minerals in the soil, preventing the movement of these chemicals into surface water or groundwater and the subsequent degradation of these waters from agricultural activity.

3.1.2 Biochar as soil conditioner

From the agricultural point of view, the application of carbonization products for soil amelioration seems to be beneficial because the treatment improves the conditions for plant growth, leading to a better yield [21]. Furthermore, due to the rapid effects and relatively low costs of such treatment, biochars are more and more frequently used in processes of soil remediation and conservation. Moreover, the application of biochars to soil leads to increased contents not only of carbon but also of other biogenic compounds, such as phosphorus, potassium, magnesium and nitrogen [7, 22]. By increasing NH_3 and NH_4^+ retention, reducing N_2O emissions and eluting NO_3 ions, as well as inducing the development of nitrogen bacteria which directly affects the increase in soil productivity, biochar helps to store nitrogen. Biochar also impacts the physical properties of soil by improving its water retention, capacity to form aggregates, and resistance to erosion. Owing to their highly porous structure, carbonization products may create favorable conditions for microorganisms, as a consequence improving the fertility and productivity of soils.

3.1.3 Improving soil for crop production

Biochar is considered to be highly effective in the restoration of the fertility of soils. Many researches confirmed that the use of biochar leads to the improvement of the soil productivity [23]. The extraordinary properties and benefits of the use of biochar are not only limited to only to the area which was disturbed for obtaining biomass to generate bio energy but it has the ability to remain persistent in the soils for almost two to three years [19]. This shows that if the biochar is applied to the lands which are not used for bio energy production, it will increase the fertility of soil and will help in reducing the pollution of soil of that land from the inorganic chemicals.

3.1.4 Nutrient availability in soils

Biochar application leads to the increase in pH of the soil and that leads to improved availability of phosphorous and potassium [9] When biochar is applied on the soil, oxidation process is observed on the surface of particles. The reason for the reported high CEC is the oxidation of aromatic carbon which leads to the formation of carboxyl groups [24]. With the increase in CEC the nutrients will remain attached to the soil opposing the leaching process. When highly oxidized organic matter attached with the surface it will create negative charge on the surface. As a result, positive charge on these sites gets decreased. However, the results from the studies showed that the effect of biochar is more expected on the soils having macro pores [25].

3.1.5 Stimulation of soil microflora and plant growth

Biochar provides a suitable habitat for a large and diverse group of soil microorganisms, although the interaction of biochar with soil microorganisms is a complex phenomenon. Addition of biochar along with phosphate solubilizing fungal strains promoted growth and yield of *Vigna radiata* and *Glycine max* plants, with better performances than control or those observed when the strains and biochar are used separately [26]. It was found that biochar increased the biological N_2 fixation (BNF) of *Phaseolus vulgaris* [27] mainly due to greater availability of micronutrients after application of biochar. It has also been reported that leaching of NH_4^+ was reduced with the application of biochar resulting to its higher availability for plant uptake [20]. Mycorrhizal fungi which were widely used as supplements for soil inoculums, often included in crop management strategies [28]. When using both biochar and mycorrhizal fungi in accordance with management practices, it is obviously possible to use potential synergism that can positively affect soil quality.

Study outline	Results summary	References
Cowpea on xanthic ferralsol	67 Mg ha^{-1} char increased biomass 150%; 135 Mg ha^{-1} char increased biomass 200%	Glaser et al. [19]
Soil fertility and nutrient retention. Cowpea was planted in pots and rice crops in lysimeters at the Embrapa Amazonia Ocidental, Manaus, Brazil	Bio-char additions significantly increased biomass production by 38 to 45% (no yield reported)	Lehmann et al. [20]
Comparison of maize yields between disused charcoal production sites and adjacent fields, Kotokosu watershed, Ghana	Grain yield 91% higher and biomass yield 44% higher on charcoal site than control.	Oguntunde et al. [29]

Table 2.
Summary of experiments assessing the impact of biochar addition on crop yield.

3.1.6 Improving crop productivity

The impactof biochar application is more prominent in highly degraded acidic or nutrient depleted soils. Several studies have reported positive responses of biochar on net primary crop production, grain yield and dry matter (**Table 2**). Little addition of charcoal (0.5 t ha^{-1}) have shown significant impact on various plant species, whereas higher rates caused plant growth inhibition [30]. Biochar if applied in combination with inorganic or organic fertilizers, can result into increased crop yields, particularly on tropical soils [19].

3.2 Environmental implications of biochar

3.2.1 Biochar and carbon sequestration

According to Turral et al. [31], agriculture generates around a fifth of the world's greenhouse gas emissions. The application of biochar is proposed as a novel approach to establish a significant, long term, sink for atmospheric carbon dioxide (CO_2) in terrestrial ecosystems. Biochar addresses two important sources of environmental problems, by sequestering CO_2 into the soil and by reducing water pollution through enhancing soil nutrient retention [32]. It was observed that biochar plays important role for emission of carbon and also essential to meeting global climate targets [33]. Biochar-bioenergy systems can play an important role in a global strategy favorably helps in carbon capture and storage at lower carbon prices whereas biochar addition to soils delivers significant increases in crop yields. Hence, effective use of biochar plays significant role in carbon sequestration.

3.2.2 Biochar and climate change

Now-a-days excessive amount of carbon dioxide is being released to the atmosphere due to the burning of fossil fuels and decomposition of biomass, which increases the carbon levels in the atmosphere day by day. Application of biochar helps in decreasing the emission of carbon dioxide as it has the ability to store 50% of the carbon from feedstock [34]. Biochar is highly stable and having the capacity to emit less carbon dioxide from organic decomposition significantly. So that it plays an important role in monitoring the release of methane and nitrogen dioxide from the soil, which are the major cause of climate change in recent days. This reduction in the release of nitrogen dioxide ensues because of the capacity of biochar to adsorb and retain the ammonium in soils and then lessen the availability of nitrogen for denitrification process. It is observed that in the fields, methane emissions were 34% higher from the fields which are treated with biochar. Though the emissions of nitrogen dioxide were found 40–51% less in soils than that of those soils which are not treated with biochar, thus global warming gases from soils decreases by amending soils with biochar [35].

3.2.3 Reducing water pollution

Application of biochar in soil also helps in the reduction of offsite pollution. It helps in increasing the retention of nutrients like phosphorous and nitrogen in soils, aid in decreasing the leaching of nutrients of soil in to the groundwater. Thus, it plays major role in saving the nutrients from erosion and nutrients availability for the cultivation of crop increases. By the pyrolysis of animal manures, a significant amount of reduction can be achieved in the mobility of phosphorous of animal manures [36] and this technique will help in reducing the weight and volume of the manures and will make the disposing off of waste easier. It also helps in conversion of the soluble inorganic phosphate present in the manure into the adsorbed phosphate in biochar.

3.2.4 Reduction of hazardous materials of environment

Biochar has the ability to sorb major environmental contaminants which are harmful for the soil. Sequestration of organic pollutants are being done by using biochar to alter their effects on the environment ultimately. Due to its struggling nature towards microorganisms and its astonishing sorption affinity, biochar acts as a critical binding phase for different organic pollutants in the environment. It was observed that the heavy metals present in the soil immensely affect the adsorption of organic pollutants on biochar and also interfere with their transport and fate. Biochar has the ability to adsorb organic contamination like persistent organic pollutants (POPs) as they have high affinity for biochar because it is naturally occurring [7].

3.3 Role of biochar on soil health

Soil health refers to the capacity of soil to perform a number of agronomic and environmental functions. Important among these functions are: agronomic/biomass productivity, response to management and inputs and resistance to biotic and abiotic stresses. With reference to agricultural land use, soil health refers to the capacity of the soil to sustain and support the growth of crops and animals while also maintaining or improving the quality of the environment. Maintaining an appropriate level of soil organic matter and biological cycling of nutrients is crucial to the success of any soil management regime. The decline in SOM contributes to several soil degradation processes including erosion, compaction, salinization, nutrient deficiency, loss of biodiversity and desertification, all of which are accompanied by a reduction in soil fertility [37]. Hence, the application of biochar and its impact on the quality of soil function is worthy of an exhaustive assessment.

According to Venkatesh et al. [38], transforming a low-value crop residue into a potentially high-value carbon source and its soil application has several important benefits. A brief review about these beneficial aspects are presented in **Table 3**.

3.3.1 Soil physical properties

Biochar as a soil amendment may improve the physicochemical properties of degraded or nutrient-depleted soils. The ability of biochar to retain soil water is a function of the combination of its porosity and surface functionality [39]. Porous

Physical properties	Chemical properties	Biological properties
• High negative charge of biochar promotes soil aggregation and structure. • Decreases bulk density, improves soil workability, reduces labour and tractor tillage and minimizing fuel emissions. • Positive effect on crop productivity by retaining plant available soil moisture due to its high surface area and porosity.	• Enhance the fertilizer use efficiency, reduce the need for more expensive fertilizers and improves the bioavailability of phosphorus and sulfur to crops. • Liming effect provides net carbon benefit compared to standard liming. • Reduce leaching of nutrients and prevents groundwater contamination. • Carbon negative process, stable carbon, longer residence period and reduces Green House Gas emissions from soil.	• High surface area, porous structure and nutrient retentive capacity of biochar provides favorable microhabitats by protecting them from drought, competition and predation. • Enhances the abundance, activity and diversity of beneficial soil bacteria, actinomycete and arbuscular mycorrhiza fungi.

Table 3.
Effect of biochar on physical, chemical and biological properties of soil.

internal structure of biochar increases soil porosity which helps to increase the surface area of soil so that water is better able to penetrate. Previous studies showed that application of biochar to infertile soils decreases soil bulk density, increases total pore volume and water holding capacity [7, 40]. Chen et al. [7] reported that biochar application decreased the tensile strength of soil cores, indicating that the use of biochar can reduce the risk of soil compaction.

3.3.2 Soil chemical properties

Biochar has potential benefits in improving the chemical properties of soils. Application of biochar to soil may improve nutrient supply to plants. Soil reaction (pH) is an important characteristic of soils in terms of nutrient availability and plant growth. Previous studies reported that soil pH was raised by high-pH biochar at about one-third the rate of lime resulting in increased calcium levels and reduced aluminum toxicity on red ferralitic soils [19, 20, 41]. Soil with a high CEC helps to hold or bind plant nutrient cations to the surface of biochar particles, humus and clay, so nutrients are retained rather than leached and therefore more available for uptake by plants [19, 20, 42]. Biochars derived from manure and animal-product feedstock are relatively rich in nutrients when compared with those derived from plant materials and especially those derived from wood [43, 44]. However, biochars in general may be more important for use as a soil amendment and driver of nutrient transformation than as a primary source of nutrients [45].

3.3.3 Soil biological sproperties

Biochar as a soil amendment is confronted with the challenge that it must benefit soil health as it can by no means be separated from soils once it is added. Soils can be viewed as complex communities of organisms that are continually changing in response to soil characteristics, climatic and management factors and especially in response to the addition of organic matter. However, compared to the addition of fresh organic matter, the addition of biochar to soils is likely to affect the diversity, abundance and activity of soil biotic communities [46]. Owing to its highly porous nature, biochar helps to provide habitat for microorganisms and also modify the biological functionality by altering the availability of substrate and activity of enzyme on, or around, biochar particles [47]. Biochar has the potential to affect microbial biomass and composition and the microbes are also able to change the properties of biochar [46]. Abujabhah et al. [48] reported that microbial abundance was improved after the addition of biochar. Biochar pores may provide physical protection for soil microorganisms. Soil reaction greatly influence microbial activity, diversity and abundance. The buffering capacity of the soil solution imparted by biochar CEC may also help to minimize pH fluctuations and maintain appropriate pH conditions in the microhabitats within biochar particles [49]. Studies have shown that biochar and fertilizer application increased microbial Biomass compared to mineral fertilizer. Microbial immobilization is an important mechanism to retain N in soils affected by leaching. Increased C availability stimulates microbial activity resulting in greater N demand, promoting immobilization and recycling of NO_3^- Biochar enhanced the PSM activity for P mobilization in phosphate rich soils and significantlyimproved the crop yield in P deficient soils [50]. The effect of biochar application on soil health under different soil types are summarized in **Table 4**.

Soil type	Biochar source	Rate of biochar addition (t ha^{-1})	Impact of biochar addition on soil health	Reference
Sandy Loam	Maize stover, Pearl millet stalk Rice and Wheat straw	20	• Maize biochar intensified soil available N and P. • Wheat biochar increased soil available K. • Rice biochar being relatively labile in soil enhanced the proliferation of microbial biomass.	Purakayastha et al. [51]
Sandy	Green cuttings	1, 10 and 40	• Increased CEC, total N and available P and Kwith biochar addition of 10 t ha^{-1}. • Increased water holding capacity of sandy soil by 6% and 25% with 10 and 40 t ha^{-1} application.	Glaser et al. [52]
Silty loam	Oak wood	7.5	• Reduced soil bulk density by 13% and increased soil-C by 7%.	Mukherjee et al. [53]
Calcareous	Rice husk and shell of cotton seed	30, 60 and 90	• Decreased soil bulk density, increased exchangeable K and water holding capacity at90 t ha^{-1}.	Liang et al. [54]
Clayey	Woody shrubs	5 and 10	• Decreased bulk density and improved saturated hydraulic conductivity as well as air capacity with 10 t ha^{-1} application.	Obia et al. [55]
Brown forest soil	Peanut shells	2.4	• Improved soil bacterial diversity, improved soil structure, increased soil pH and promoted effectiveness of soil nutrients.	Wang et al. [56]
Anthrosol	Wheat straw	10 and 40	• SOC increased by 57%, total N content enhanced by 28% in the 40 t ha^{-1}.	Afeng et al. [57]

Table 4.
Summary of the effect of biochar additions on soil health under different soil types.

4. Limitation

It is complicated to speculate the agronomic effectiveness of biochar with limited studies being conducted in different soil types, climatic belts and land use conditions. Heterogeneous nature as well as production cost of biochar for research and field application will continue to remain a major limitation until commercial-scale pyrolysis facilities are established. Some of the experimental constraints on use of biochar in agricultural systems are [38, 58]:

i. Unavailability of sufficient quantity of biochar for large scale use

ii. Susceptibility of dry biochar to wind erosion

iii. Non-adoption of biochar by local farming communities

iv. Unavailability of adequate farm labour

v. Expensive wage costs incurred for collection and processing of crop residues

vi. Lack of appropriate farm machinery for on-farm recycling of crop residues

vii. Inadequate incentives for recycling of crop residue.

Other limitations involve contamination risk of biochar (PAHs, heavy metals, dioxins) when contaminated feedstocks are used or when inappropriate process conditions are used for biochar preparation such as temperature greater than 500°C.Removal of crop residues from the field for biochar production results in its reduced incorporation into soils, hampering many soil properties. In certain cases, extremely high rate of biochar application produces negative effects on earthworm survival rates [59].

5. Recommended practices for use of biochar in agriculture

Nature of crop residues for biochar production: Freshly harvested, under-utilized dry crop and agroforestry residues should be used for biochar production. Use of crop residues grown on toxic chemical and heavy metal contaminated site should be avoided.

Location and operation of biochar kiln unit: The biochar kiln unit should be located near to the crop and agroforestry residue generating locations to providea management solution and minimize handling and transportation loss and costs. The kiln unit should be operated in an open space with sufficient atmospheric aircirculation, ideally away from any other structures. Precautions should be maintained while opening and closing of kilning unit during the cooling period.

Preparation of biochar: The fresh biochar needs to be 'cured' overnight with exposure to open air. It is advisable to store the biochar outside under shelter, away from buildings, in a cool, dry wellventilatedopen spot and grind to powder just before its use. The biochar should be transportedto the application site in a sealed container or in a closed plastic bag.

Application of biochar: It is better to apply biochar as close to ground as possible on mildwindy day to avoid drift loss by wind.

Protective clothing such as insulated gloves or gunny rags, masks or cloths should be used whenever possible while handling kiln and biochar.

6. Future prospects

Although biochar utilization has gained much attention in recent years, there still lies a huge knowledge gap that needs to be addressed. The ultimate fate of biochar under field conditions and its long-term influenceon soil quality are questions which remains unanswered. The influence of biochar on soil physical and chemical properties as well as the microbial communities, needs to be further explored especially in accordance to changes in biogeochemical cycles [60]. Researchers need to searchsolutions to reduce GHG emission to a large extent, when soils are amended with biochar [61]. Focus needs to be paid on full-scale outdoor trials of biochar as a way to restore contaminated soils and evaluate how long biochar retains the metals as it ages in the field [15]. Lastly, better understanding of biochar preparation needs to be done with different feedstock materials and pyrolysis processes to target specific soil deficiencies [60].

7. Conclusion

Ever-increasing population has paved the way to agricultural land depletion which needs to be controlled by adopting sustainable crop production practices. Direct incorporation of crop residues into agricultural soil conserves soil nutrients as well as organic carbon contentbut initiates considerable crop management problems by delayingthe decomposition process. Conversion of crop residues to biochar by thermo-chemical process (slow pyrolysis) is an attractive, economical alternative approach for effective management and disposal of these excess crop residues, whichotherwise are being used inefficiently. Addition of biochar to soil is one of the best practices to overcome any biotic/abiotic stress caused, such as heavy metal toxicity, soil acidity, nutrient unavailability etc. and to increase the crop productivity. From the agriculture point of view, application of biochar as a soil conditioner generates numerous benefits, such as improvement of the physical, chemical and biological properties of soils, and this in turn contributes to an increased crop yield. Owing to their physicochemical properties, biochar can be used for soil carbon sequestration, reduction of the bioavailability of contaminants affecting living organisms as well as for water treatment. The persistence of biochar effects on soil processes and mechanisms remains to be resolved under realistic field conditions. Therefore, it is recommended to use biochar as a soil amendment for enhancing soil health and environmental condition as well as long-term carbon sequestration.

DOI: http://dx.doi.org/10.5772/intechopen.97136

Author details

Shreya Das[1], Samanyita Mohanty[1], Gayatri Sahu[2]*, Mausami Rana[1] and Kiran Pilli[1]

1 Bidhan Chandra Krishi Viswavidyalaya, Mohanpur, Nadia, West Bengal, India

2 Siksha 'O' Anusandhan Deemed to be University, Bhubaneswar, Odisha, India

*Address all correspondence to: gayatribckv@gmail.com

References

[1] Pullagurala VL, Rawat S, Adisa IO, Hernandez-Viezcas JA, Peralta-Videa JR, Gardea-Torresdey JL (2018) Plant uptake and translocation of contaminants of emerging concern in soil. Science of the Total Environment 636:1585-1596.

[2] Samantarai SK and Achakzai AKK (2014) Application of nanotechnology in agriculture and food production: opportunity and challenges. Middle East Journal of Scientific Research 22(4): 499-501.

[3] Sandhu SS, Ussiri DA, Kumar S, Chintala R, Papiernik SK, Malo DD, Schumacher TE (2017) Analyzing the impacts of three types of biochar on soil carbon fractions and physiochemical properties in a corn-soybean rotation. Chemosphere 184:473-481.

[4] Sun Z, Sänger A, Rebensburg P, Lentzsch P, Wirth S, Kaupenjohann M, Meyer-Aurich A (2017) Contrasting effects of biochar on N2O emission and N uptake at different N fertilizer levels on a temperate sandy loam. Science of the Total Environment 578:557-565.

[5] Bashir S, Zhu J, Fu Q, Hu H (2018) Cadmium mobility, uptake and anti-oxidative response of water spinach (Ipomoea aquatic) under rice straw biochar, zeolite and rock phosphate as amendments. Chemosphere 194:579-587.

[6] Kiran YK, Barkat A, Cui XQ, Ying FE, Pan FS, Lin TA, Yang XE (2017) Cow manure and cow manure-derived biochar application as a soil amendment for reducing cadmium availability and accumulation by Brassica chinensis L. in acidic red soil. Journal of Integrative Agriculture 16(3):725-734.

[7] Chen J, Zhu D, Sun C (2007) Effect of heavy metals on the sorption of hydrophobic organic compounds to wood charcoal. Environmental Science & Technology 41(7):2536-2541.

[8] Venkatesh G, Gopinath KA, Reddy KS, Reddy BS, Prasad JV, Rao GR, Pratibha G, Srinivasarao C, Chary GR, Prabhakar M, Kumari VV (2018) Biochar for enhancing the soil, water and crop productivity in rainfed areas. In: Krishna Rao B, Sammi Reddy K, Visha Kumari V, Nagarjuna kumar R and Rejani R. (eds.) Rainwater Management for Climate Resilient Agriculture in Drylands. pp 523.

[9] Atkinson CJ, Fitzgerald JD, Hipps NA (2010) Potential mechanisms for achieving agricultural benefits from biochar application to temperate soils: a review. Plant and soil 337(1-2):1-8.

[10] Li L, Rowbotham JS, Greenwell CH, Dyer PW (2013) An introduction to pyrolysis and catalytic pyrolysis: versatile techniques for biomass conversion. Elsevier:173-208.

[11] Zhang A, Cui L, Pan G, Li L, Hussain Q, Zhang X, Zheng J, Crowley D (2010a) Effect of biochar amendment on yield and methane and nitrous oxide emissions from a rice paddy from Tai Lake plain, China. Agriculture, ecosystems & environment 139(4): 469-475.

[12] Zhang L, Xu CC, Champagne P (2010b) Overview of recent advances in thermo-chemical conversion of biomass. Energy conversion and management 51(5):969-982.

[13] Britt PF, Buchanan AC, Cooney MJ, Martineau DR (2000) Flash vacuum pyrolysis of methoxy-substituted lignin model compounds. The Journal of organic chemistry 65(5):1376-1389.

[14] Liu Y, Leahy JJ, Grams J, Kwapinski W (2019) Hydro-Pyrolysis and Catalytic

Upgrading of Biomass and Its Hydroxy Residue Fast Pyrolysis Vapors. Energies. 12(18):3474.

[15] Zhang M, Cheng G, Feng H, Sun B, Zhao Y, Chen H, Chen J, Dyck M, Wang X, Zhang J, Zhang A (2017) Effects of straw and biochar amendments on aggregate stability, soil organic carbon, and enzyme activities in the Loess Plateau, China. Environmental Science and Pollution Research 24(11):10108- 10120. https://doi.org/10.1007/s11356-017-8505-8.

[16] Kavitha B, Reddy PV, Kim B, Lee SS, Pandey SK, Kim KH (2018) Benefits and limitations of biochar amendment in agricultural soils: A review. Journal of environmental management 227:146-154.

[17] Hammes K, Schmidt MW (2009) Changes of biochar in soil. In: Lehmann M, Joseph S (ed) Biochar for environmental management: Science and technology. Earthscan, London, pp. 169-182.

[18] Mensah AK, Frimpong KA (2018) Biochar and/or compost applications improve soil properties, growth, and yield of maize grown in acidic rainforest and coastal savannah soils in Ghana. International Journal of Agronomy 1:1-8.

[19] Glaser B, Lehmann J, Zech W (2002) Ameliorating physical and chemical properties of highly weathered soils in the tropics with charcoal–a review. Biology and fertility of soils 35(4):219-230.

[20] Lehmann J, da Silva JP, Steiner C, Nehls T, Zech W, Glaser B (2003) Nutrient availability and leaching in an archaeological Anthrosol and a Ferralsol of the Central Amazon basin: fertilizer, manure and charcoal amendments. Plant and soil 249(2):343-357.

[21] Macdonald LM, Farrell M, Van Zwieten L, Krull ES (2014) Plant growth responses to biochar addition: an Australian soils perspective. Biology and fertility of soils 50(7):1035-1045.

[22] Laird DA (2008) The charcoal vision: a win–win–win scenario for simultaneously producing bioenergy, permanently sequestering carbon, while improving soil and water quality. Agronomy Journal 100(1):178-181.

[23] Lehmann J, Rondon M (2006) Biochar soil management on highly weathered soils in the humid tropics. In: Uphoff N (ed.), Biological approaches to sustainable soil systems. Boca Raton, FL: CRC Press, pp. 517-530.

[24] Liang B, Lehmann J, Solomon D, Kinyangi J, Grossman J, O'neill B, Skjemstad JO, Thies J, Luizão FJ, Petersen J, Neves EG (2006) Black carbon increases cation exchange capacity in soils. Soil Science Society of America Journal 70(5):1719-1730.

[25] Tryon EH (1948) Effect of charcoal on certain physical, chemical, and biological properties of forest soils. Ecological Monographs 18(1):81-115.

[26] Saxena J, Rawat J, Kumar R (2017) Conversion of biomass waste into biochar and the effect on mung bean crop production. CLEAN –Soil, Air, Water 45(7):1-9.

[27] Rondon M, Lehmann J, Ramirez J, Hurtado M (2007) Biological nitrogen fixation by common beans (*Phaseolus vulgaris* L.) increases with bio-char additions Biology and Fertility of Soils 43:699-708.

[28] Schwartz MW, Hoeksema JD, Gehring CA, Johnson NC, Klironomos JN, Abbott LK, Pringle A (2006) The promise and the potential consequences of the global transport of mycorrhizal fungal inoculum. Ecology letters 9(5):501-515.

[29] Oguntunde PG, Fosu M, Ajayi AE, Van De Giesen N (2004) Effects of

charcoal production on maize yield, chemical properties and texture of soil. Biology and Fertility of Soils 39(4):295-299.

[30] Ogawa M, Okimori Y, Takahashi F (2006) Carbon sequestration by carbonization of biomass and forestation: three case studies. Mitigation and adaptation strategies for global change 11(2):429-444.

[31] Turral H, Burke J, Faurès JM (2011) Climate change, water and food security. Food and Agriculture Organization of the United Nations (FAO), Rome, Italy.

[32] Shackley S, Sohi S, Brownsort P, Carter S, Cook J, Cunningham C, Gaunt J, Hammond J, Ibarrola R, Mašek O, Sims K (2010) An assessment of the benefits and issues associated with the application of biochar to soil. Department for Environment, Food and Rural Affairs, UK Government, London.

[33] Woolf D, Lehmann J, Lee DR (2016) Optimal bioenergy power generation for climate change mitigation with or without carbon sequestration. Nature Communications 7(1):1-11.

[34] Sohi SP, Krull E, Lopez-Capel E, Bol R (2010) A review of biochar and its use and function in soil. In: Advances in agronomy. Academic Press, pp. 47-82.

[35] Zwieten L, Singh B, Joseph S, Kimber S, Cowie A, Chan KY (2009) Biochar and emissions of non-CO2 greenhouse gases from soil. In: J. Lehmann, S. Joseph (ed) Biochar for Environmental Management: Science and Technology. Earthscan, London, pp. 227-249.

[36] He BJ, Zhang Y, Funk TL, Riskowski GL, Yin Y (2000) Thermochemical conversion of swine manure: an alternative process for waste treatment and renewable energy production. Transactions of the ASAE 43(6): 1827-1833.

[37] Lal R (2015) Sequestering carbon and increasing productivity by conservation agriculture. Journal of Soil and Water Conservation 70(3):55A-62A.

[38] Venkatesh G, Gopinath KA, Sammi Reddy K (2015) Low-cost portable kiln for biochar production from on-farm crop residue. Indian farming 64(12):9-12.

[39] Suliman W, Harsh JB, Abu-Lail NI, Fortuna AM, Dallmeyer I, Garcia-Pérez M (2017) The role of biochar porosity and surface functionality in augmenting hydrologic properties of a sandy soil. Science of the Total Environment 574:139-147.

[40] Abel S, Peters A, Trinks S, Schonsky H, Facklam M, Wessolek G (2013) Impact of biochar and hydrochar addition on water retention and water repellency of sandy soil. Geoderma 202:183-191.

[41] Steiner C, Teixeira WG, Lehmann J, Nehls T, de Macêdo JL, Blum WE, Zech W (2007) Long term effects of manure, charcoal and mineral fertilization on crop production and fertility on a highly weathered Central Amazonian upland soil. Plant and soil 291(1-2):275-290.

[42] Laird D, Fleming P, Wang B, Horton R, Karlen D (2010) Biochar impact on nutrient leaching from a Midwestern agricultural soil. Geoderma 158(3-4):436-442.

[43] Alburquerque JA, Calero JM, Barrón V, Torrent J, del Campillo MC, Gallardo A, Villar R (2014) Effects of biochars produced from different feedstocks on soil properties and sunflower growth. Journal of Plant Nutrition and Soil Science 177(1):16-25.

[44] Singh B, Singh BP, Cowie AL (2010) Characterisation and evaluation of biochars for their application as a soil amendment. Soil Research 48(7):516-525

[45] DeLuca TH, Gundale MJ, MacKenzie MD, Jones DL (2015) Biochar effects on soil nutrient transformations. Biochar for environmental management: science, technology and implementation 2:421-454.

[46] Lehmann J, Rillig MC, Thies J, Masiello CA, Hockaday WC, Crowley D (2011) Biochar effects on soil biota–a review. Soil biology and biochemistry 43(9):1812-1836.

[47] Gomez JD, Denef K, Stewart CE, Zheng J, Cotrufo MF (2014) Biochar addition rate influences soil microbial abundance and activity in temperate soils. European Journal of Soil Science 65(1):28-39.

[48] Abujabhah IS, Bound SA, Doyle R, Bowman JP (2016) Effects of biochar and compost amendments on soil physico-chemical properties and the total community within a temperate agricultural soil. Applied Soil Ecology 98:243-253.

[49] Rousk J, Bååth E, Brookes PC, Lauber CL, Lozupone C, Caporaso JG, Knight R, Fierer N (2010) Soil bacterial and fungal communities across a pH gradient in an arable soil. The ISME journal 4(10):1340-1351.

[50] Deb D, Kloft M, Lässig J, Walsh S (2016) Variable effects of biochar and P solubilizing microbes on crop productivity in different soil conditions. Agroecology and Sustainable Food Systems 40(2):145-168.

[51] Purakayastha TJ, Kumari S, Pathak H (2015) Characterisation, stability, and microbial effects of four biochars produced from crop residues. Geoderma 239:293-303.

[52] Glaser B, Wiedner K, Seelig S, Schmidt HP, Gerber H (2014) Biochar organic fertilizers from natural resources as substitute for mineral fertilizers. Agron Sustain Dev 35: 667-678. https:// doi.org/10.1007/ s13593-014-0251-4.

[53] Mukherjee A, Lal R, Zimmerman AR (2014) Effects of biochar and other amendments on the physical properties and greenhouse gas emissions of an artificially degraded soil. Science of the Total Environment 487:26-36.

[54] Liang F, Li GT, Lin QM, Zhao XR (2014) Crop yield and soil properties in the first 3 years after biochar application to a calcareous soil. Journal of Integrative Agriculture 13(3):525-532.

[55] Obia A, Mulder J, Hale SE, Nurida NL, Cornelissen G (2018) The potential of biochar in improving drainage, aeration and maize yields in heavy clay soils. PLoS ONE 13(5): e0196794.

[56] Wang H, Ren T, Feng Y, Liu K, Feng H, Liu G, Shi H (2020) Effects of the Application of Biochar in Four Typical Agricultural Soils in China. Agronomy 10(3):351.

[57] Afeng Z, Liqiang C, Gengxing P, Lianqing L , Qaiser H, Xuhui Z, Jinwei Z, David C (2010) Effect of biochar amendment on yield and methane and nitrous oxide emissions from a rice paddy from Tai Lake plain, China. Agriculture, Ecosystems and Environment 139: 469-475.

[58] Parmar A, Nema PK, Agarwal T (2014) Biochar production from agro-food industry residues: a sustainable approach for soil and environmental management. Current Science 107(10):1673-1682.

[59] Liesch AM, Weyers SL, Gaskin JW, Das KC, Gilkes RJ, Prakpongkep N (2010) Impact of two different biochars on earthworm growth and survival. Annals of Environmental Science 4:1-9.

[60] Ding Y, Liu Y, Liu S, Li Z, Tan X, Huang X, Zeng G, Zhou L, Zheng B

(2016) Biochar to improve soil fertility. A review. Agronomy for sustainable development 36(2):36.

[61] Mandal S, Sarkar B, Bolan N, Novak J, Ok YS, Van Zwieten L, Singh BP, Kirkham MB, Choppala G, Spokas K, Naidu R (2016) Designing advanced biochar products for maximizing greenhouse gas mitigation potential. Critical Reviews in Environmental Science and Technology 46(17):1367-1401.